BREAKING BAD HABITS IN HORSES

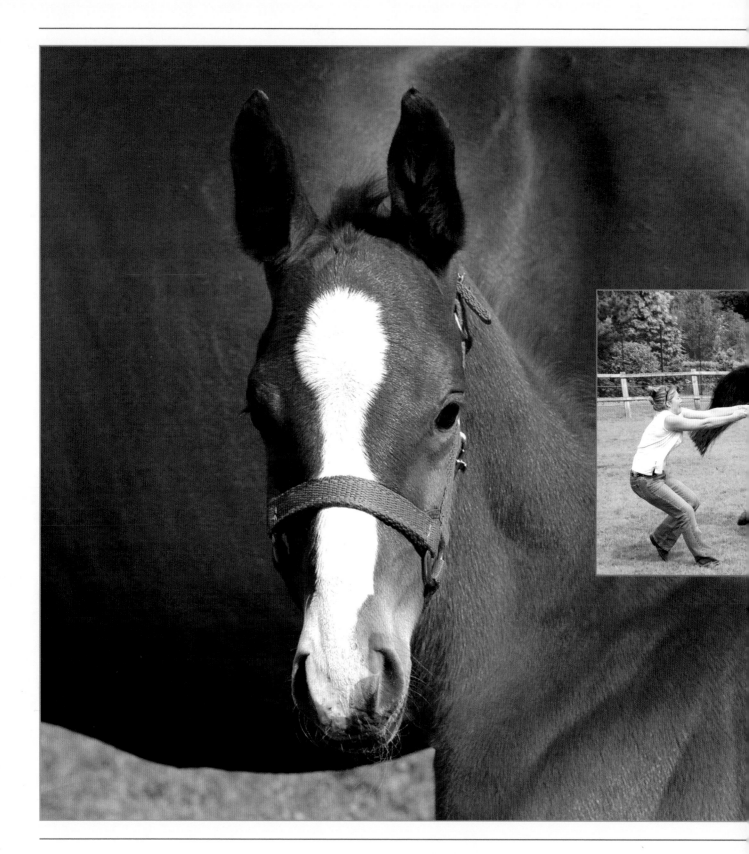

BREAKING
BAD HABITS
IN
HORSES

Tried-and-tested methods of remedying faults and problem behaviours in horses

JO BIRD

Interpet Publishing

Published by **Interpet Publishing**,
Vincent Lane, Dorking,
Surrey RH4 3YX, England

ISBN 978-1-84286-137-0

Editor: **Philip de Ste. Croix**
Designer: **Philip Clucas** MCDS
Photographer: **Neil Sutherland**
Diagram artwork: **Maggie Raynor**
Index: **Richard O'Neill**
Production management:
Consortium, Suffolk
Print production: **SNP Leefung,China**

Acknowledgements
Thanks to all at Sworders and my
family for putting up with my incessant
horse talk. Thank you to my models
especially those that had to perform
the 'problems': Laura Key, Gill and
Hannah Leage, Viv Perry, Jenny Dear,
Sarah Lee, Adele Rawlinson, Lucy and
Kate Fallen, Steph Kerans, Jess
MacKenzie-Lambert, Gill Walker, Claire
Mardell, Sam Jolliffe, Katy Griffiths and
Ian Mason. Thanks also to Yvonne
Smith and Katy Griffiths for providing
props. Finally, thank you to Philip,
Malcolm, Neil and Kevin – we seem to
gel pretty well as a team now!

Jo Bird

Jo Bird has owned horses for most of her life and used to juggle working as a groom in the mornings and going to an office job in the afternoons! She now provides management and nutritional advice to people buying horses and has worked in an advisory capacity helping to develop new products for a leading equestrian product manufacturer. She has owned a variety of horses, from foals to aged veterans and from huge, heavy traditional cobs to fine, fit performance horses. She now owns several racehorses and is on the road to obtaining her Trainer's Licence. Her natural horse-manship philosophy is 'Think about it from the horse's point of view'. She is the author of *Keeping A Horse The Natural Way*.

■ CONTENTS

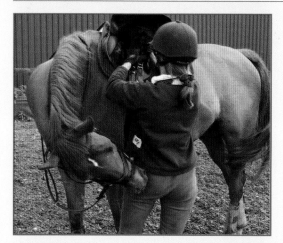

■ BEFORE YOU START – ESSENTIAL ADVICE

Health and Fitness

We all have off days, and horses do too. They may or may not be suffering from a virus or an ailment that actually requires a visit from the vet, but it makes sense to conduct a quick top-to-tail health score before starting any ridden work. Even if the visual score is 100 per cent, if you subsequently feel that your horse is not performing well, this could be due to tiredness or aching muscles or he could simply be in the wrong frame of mind. If this is the case, there is little point in battling to make him work correctly. It may be better to make an allowance and leave it for that day.

Similarly, if you have had a bad day and are wound up, stressed or tired, it is better to go for a walk (or a stiff drink!) than potentially vent your frustration or blow a short fuse on an unsuspecting horse who makes an error.

Safety

We are all aware that horses are unpredictable by nature. To own a 'bombproof' calm horse is a blessing as your chances of getting hurt are lessened, although by no means entirely eliminated. A horse with a known 'vice' of course simply increases the risks, but **never, ever be complacent** with any horse.

Check General Health and Demeanour

Is he alert and attentive and pleased to see you?	**NO** / **YES**	
Is there any discharge from his muzzle or does he have dull, weepy eyes?	**YES** / **NO**	
Are his eyes bright and clear?	**NO** / **YES**	
Does he seem sullen and disinterested or bad-tempered towards you?	**YES** / **NO**	
Are the mucous membranes around his eyes, nostrils and gums a salmon-pink colour?	**NO** / **YES**	
Does he seem agitated, uncomfortable or distressed?	**YES** / **NO**	
Does his posture indicate he is sound?	**NO** / **YES**	
Is he shifting his weight, resting his legs, head-shaking or generally looking unsettled or uncomfortable?	**YES** / **NO**	

■ Do not think of riding until you have investigated the cause of the problem

✔ He appears to be healthy and happy, so continue

☐ **Check body visually.** Check body and limbs for cuts, punctures, wounds, bites, sores, girth galls, over-reach grazes, mud fever etc. Check legs are not puffy or swollen. Pick out hooves and check for excessive heat in them.

☐ **Check tack.** Check the fit of all tack and its positioning. Make sure all surfaces in contact with the skin are free of grease and mud which could cause chafing.

☐ **Tighten girth straps gradually and check bit.** Check position of bit in the mouth. Does the horse seem comfortable with it or could there be a problem? Check for sores or cuts in the mouth.

☐ **Flex the horse before mounting.** Stretch the forelegs to eliminate pinching from the girth. Lead the horse in a small circle, to the left and to the right to allow him to feel the tack and loosen up.

☐ **Mount slowly.** Ideally mount from a mounting block and lower your weight gently into the saddle.

☐ **Warm up slowly.** Begin riding at a walk for a minimum of five minutes. Start on a loose rein to allow him to stretch out. He should be happy and interested in his work.

Left: Be safe, be seen – a hard hat plus fluorescent clothing or horse accessories will improve safety especially when riding on the roads.

A horse rider's wardrobe can be very diverse in style, but the essential components which should be worn for safety reasons are:

Protective Hat or Helmet

When selecting a hat or helmet the main criteria are that it fits well and does the job of protecting your head. It is not only when you fall off that helmets are invaluable – many a time my hat has saved me from branches springing back in my face or brambles catching in my hair in a wooded area. It needs to be up to the current safety standards BS EN1384, PAS015 or ASTM F1163. Not only does this ensure that you will have the optimum protection but you will also find that most shows and events will not even allow you to ride without a hat meeting current guidelines. The wide array of hats and helmets on the market means there is plenty to choose from.

Below: Safety hats and clothing are designed for comfort and protection from adverse weather, as well as impact resistance in the event of a fall.

Both above: Whatever style you choose, footwear should be sturdy (ideally with steel toe caps) with a small heel and not be too wide for the stirrup iron.

Gloves

Gloves are useful for keeping a secure grip on the reins while riding and they also protect you from another hazard of horsekeeping which is often ignored – that of leading a horse. Frequently during lungeing or while being led to or from the field, horses may try to escape and drag their handlers. Loading and unloading a horse onto a lorry or trailer can also be dangerous if the horse rushes out. I have seen horribly rope-burned hands and a woman in tears as she tried to soothe the agony of raw skin by keeping her hands in a bucket of cold water. It pays to wear gloves.

Boots

Again, styles range from short rubber mucking-out boots to long rubber or leather boots, cowboy boots and fashion boots. Depending whether you are working around horses on the ground or the style of riding or the formality of the occasion these can be worn with jeans or jodhpurs and with or without chaps (leather or fabric calf protectors). Although they are generally heavier, a steel toecap is also recommended, especially if you are working with young or boisterous horses who have not learned to keep out of your 'personal zone'.

In addition, I always recommend the following:

continued ➡

■ BEFORE YOU START – ESSENTIAL HEALTH & SAFETY

Body Protector

These should not be seen as purely for use by event or endurance riders –anyone with even a slightly 'spooky' horse should not go out without wearing one. They protect your spine and ribcage not only from the impact

Left: Modern body protectors are lightweight and quite comfortable.

of a fall but also against the flailing hooves of a horse struggling to get up, perhaps from a ditch into which you have both fallen.

If you are put off by the weight or the feeling of restriction, bear in mind that the latest materials technology is improving the design of these every year. For medium protection a body protector designed for jockeys offers a featherlight option.

Fluorescent Clothing

There is a huge selection of fluorescent items for both horse and rider so you can opt for wearing a tabard or hat cover on a dull evening or use fluorescent sleeving or reflective strips on

---

Safety in the Stable Yard and Paddocks

Stables/Stalls
Check the woodwork for splinters and that any metal work is not bent. Run a wet sponge round the stable from floor to ceiling – not only will this clean away the dust and cobwebs but it will catch on any protruding nails which may not be obvious to the eye in such a dark environment but which could cut and scar your horse.

Yard
Make sure all visitors know that there is 'No Smoking' in the yard. Horse hay and bedding such as straw, paper, woodchip or hemp are highly flammable. Keep fire extinguishers in prominent places and ensure you know how to use them.

Use Perspex rather than glass in any windows and try not to bring any glasses or bottles onto the yard where they could easily be knocked over and break.

Paddocks
Regularly inspect fencing for broken rails or sagging wire. Gates should be re-hung if they drag on the ground. Have the fields rolled if poaching has caused holes and rutting which could cause a twisted ankle or a lame horse.

Muck
Store muck safely so that there is no run-off into water sources. Keeping it away from the yard will reduce the fly problem.

Rules
Have some Yard Rules posted on a notice board in a prominent position. These do not need to be too onerous but should concentrate on those that relate to horse and rider welfare including:

- Always tie a horse up to a piece of baler twine and never directly to the fence or tie-ring. This will break in an emergency if a horse panics and so prevent it from injuring itself as it struggles to get free.
- Carry a mobile phone at all times.
- Wear appropriate clothing.
- No smoking.

Above: However beautiful you may look, this type of shoe is never compatible with horseshoes! Sturdy footwear is necessary at all times when you are around horses.

Make sure the telephone contact numbers of family and vets relating to you and other people are easily obtained in an emergency.

Health and Safety Tips

the horse. Stirrup lights are also available if needed but bear in mind that if visibility is poor, then you should think carefully about whether it is safe to go out at all.

Horsewear
- Check your tack each week for rips or weaknesses, especially around buckles.
- Clean and oil your tack regularly to protect it.
- Rubber grip reins are useful in winter as they do not slip through your fingers as wet leather can do.
- Check the condition of your horse's feet and, if shod, that the shoes are firm and unlikely to come loose during a ride which could be dangerous.
- Boots and bandages – only use these if essential. In my view, boots and bandages should only be used to protect from exceptional exertion causing strains and not for your normal work. Regular use of boots and bandages causes a horse to

be reliant on their support and can cause problems from sweat, damp or dirt being trapped underneath or general rubbing causing sores.

Remember, if there is any doubt in your mind – don't risk it! Never take chances, especially if you are on your own. Wait until you are in a controlled environment with experienced people to help you. Keeping and riding horses should be a pleasurable pastime and this is only achievable if you think calmly and logically about horse handling and try to anticipate and pre-empt any potential pitfalls.

Safety on the Road or Trail
Always make sure you take a mobile phone so you can contact someone for help if necessary in an emergency, and ensure that you are visible if the light is failing. Do not attempt to take a nervous horse out on roads alone – make sure you go out with someone else on an unflappable horse or lead the inexperienced one at the side of the calm one who is nearest the traffic. Know your road signals and stay alert (not engrossed chatting to your friend!). Have insurance to cover yourself, your horse plus third party liability and if possible take your British Horse Society Riding & Road Safety Test.

Above left: Keep a fire extinguisher in a prominent place on the stable yard and ensure that it is serviced annually and that all persons are instructed how to use it.

Above right: Bottles and cups brought onto the yard should be made of plastic only and be disposed of safely after use. Glass or china containers could shatter.

Right: Even on a 'safe' horse, a hat is invaluable on a trail to protect from overhanging branches or brambles which could catch in the rider's face or hair.

CHAPTER ONE

BEHAVIOURAL AND HANDLING PROBLEMS

When you embark on any activity involving horses, there is certainly a learning curve to surmount. Horses have their own complex language which allows an established herd to live together and communicate in a way which normally only involves aggression as a last resort. In a domestic situation things are more unsettled and by 'horse watching' and taking the time to understand their complex natures, we can improve our partnerships with them and so make handling them very much easier.

■ HERD DISHARMONY/POOR SOCIAL SKILLS

■ Problem

Above: Male horses will naturally fight for supremacy and even mares and geldings in a domestic situation will establish a pecking order through threats and physical confrontations.

In a domestic situation horses are usually thrown together both for our convenience and, of course, for their well-being as this provides companionship. For the most part, this works well. There are times, however, when aggressive behaviour between animals can rear its head. This can be terrifying to watch and no owner wants one of their animals causing upset or injury to other horses. The subservient animals can be traumatized by bullying and I have had one of my own animals suffer fits as a direct result of this stress. One kick can result in an animal having to be destroyed or its intended competitive career rendered useless and, of course, there is the considerable expense of vet's bills to consider on top of this.

■ Cause

In the majority of cases, horses and ponies on a yard are not related, nor will they necessarily have been kept together for a long period of time. Horses are also sold and replaced regularly. This turnover in the membership of a herd is very different from the manner in which a herd would regenerate and change in its natural state. In the wild, the group would remain far more constant with just new births increasing numbers while maturing colts would leave to form their own herds and elderly or injured animals would die or get left behind. The central core of the family, however, would remain unchanged throughout.

During their lives, wild horses experience these four main types of relationship:

1 Stallion + mare *(above)* – procreation and future viability of the herd.
2 Mare + foal *(below left)* – maternal bonding.

3 Inter-herd relationships *(above right)* – hierarchy and reliance on group stability.
4 Pair-bonds *(below)* – particularly close peer attachments between two animals.

Behavioural and Handling Problems

Pair-bonds affect herd status

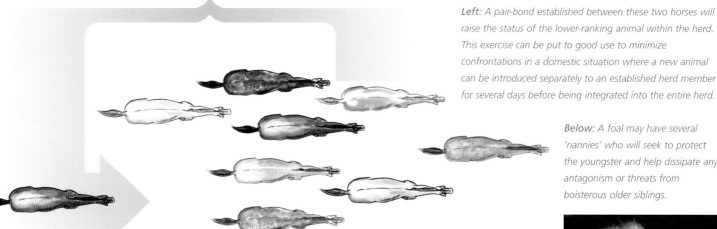

Left: A pair-bond established between these two horses will raise the status of the lower-ranking animal within the herd. This exercise can be put to good use to minimize confrontations in a domestic situation where a new animal can be introduced separately to an established herd member for several days before being integrated into the entire herd.

Below: A foal may have several 'nannies' who will seek to protect the youngster and help dissipate any antagonism or threats from boisterous older siblings.

In domestic herds it is rare for horses to experience more than three of these relationships and, more usually, although they will be part of a herd (however fluid its nature), they will not often enjoy a pair-bond that lasts more than a couple of years. Stallions are rarely allowed to run with the mares and much modern breeding now consists of artificial insemination so the horses don't need to get together even once!

Problems can arise when foals are kept only with their dam or alternatively are weaned too early. Just like children, much of their herd behaviour and social skills is learned from their peers and elders and it is this formative education that moulds how they will interact in later life.

Herd rank is something that changes and evolves depending on the individual horses in the group. A horse's status is not dictated by its size, sex or length of time in the group. It is often amusing to watch how the largest, 'bolshie' horses can be put in their place by the introduction of a new horse which turns the pecking order upside down. Rank is maintained by physical displays of intimidation, pressure and threats. In the wild the motivation to achieve a high status is powerful as a superior position helps in the survival stakes. The very young and the very old are always likely to be lower ranking due to their vulnerability and lack of physical strength, but foals are likely to be under the protection of a higher ranking adult who will to help fight their battles. Often an adult mare will be the ultimate boss and in the wild she would dictate the route the herd would follow and be allowed to drink first at any water source.

Once status is established (with members at various levels of ranking) the herd becomes harmonious and few aggressive altercations take place. A lower-ranking horse may form a pair-bond with a higher ranking horse and will not fear him. Pair-bonds share obvious deep-rooted understanding and co-dependency and will call out desperately and fence walk when separated, but show little interest if another herd member is taken away.

Remedy ➡

■ HERD DISHARMONY/POOR SOCIAL SKILLS

■ Remedy

It is unusual for the average owner to have a problem with colts and stallions striving for authority within a herd. Most male horses are gelded; however, many people like to separate mares from geldings within livery establishments to limit rivalry, as some gelded horses can display 'riggy'* behaviour and stir up the mares who may well lash out.

If you have a horse that is either scared of the other horses (evidenced by standing alone on the other side of the paddock or hanging back if hay is put down etc.) or alternatively is overly domineering and the troublemaker in the group, remove them and select a pair-bond for them.

Above: Keeping two horses separate from the rest means they are reliant on one another for company and often aggressive or submissive behaviour will subside.

Above: If you notice that your horse is regularly excluded from the herd, he may be fearful which can manifest itself as illness or in behavioural problems.

The most important criterion for choosing a pair-bond is that hopefully they will be together for a long time. There are many rescue centres desperate for fosterers for 'companion

only' horses and ponies and they can become invaluable little treasures. A pair-bond can be of either sex but if you know your gelding will incessantly chase a mare, then it would be sensible to opt for another gelding. Often older, retired animals will be more phlegmatic and tolerant and will teach confidence and social skills to the younger partner without aggressive challenges.

Section off an area of the field if you do not have a separate paddock but the bonding will establish faster if they are kept totally away from the rest of the animals. You will know that the bonding is becoming established if they groom each other or lie down together (or one lies down and the other watches over him).

Then either add other horses to their group (two at a time) or move the newly established pair-bond back into the herd. Be prepared to resign yourself to the fact that there are some

---

* A rig (who can display riggy behaviour) is a male horse or pony who is thought to have been gelded but either still maintains one or both testes undescended or who may have been gelded relatively late and has delusions of stallionhood.

*Right: Horses want to be friendly –
it is normally fear or food
competitiveness that causes
confrontations. Close companions
will allow each other into their
personal space and mutual
grooming gives both physical and
mental pleasure. By watching your
horses, you can identify the areas
from which they appear to derive
most pleasure when scratching or
itching themselves on objects.*

*Below: With electric tape it is easy
to partition off sections of a field to
provide paddocks for varying
numbers of horses. This is often
preferable than risking a free-for-all
fight, especially if the horses are
owned by different people.*

combinations that will never work and certain horses may
have to go in sub-groups of their own to avoid further
confrontations.

It is always healthier to allow horses to act in a natural and
instinctive manner but, if this is not possible, it is easy to create
sectioned-off areas (with electric tape) that allow horses to
graze close to one another but which still keeps them separated
to prevent injury.

If you take your horse anywhere where he may end up in
close proximity to other horses (e.g. in the collecting ring
before a competition, out hunting etc.) and you know that your
horse is likely to lash out, it is advisable to put a red band
(ribbon or red insulating tape will do) around the top of your
horse's tail. This acts as a warning to other riders not to bring
their horses too close but do remember that it is your
responsibility to keep your own horse under control. In such
cases, Third Party insurance is a must.

■ BULLYING/AGGRESSIVE BEHAVIOUR

■ Problem

Aggressive behaviour towards humans or other horses is a truly serious problem, especially for owners who keep their animals in public yards where they could find themselves liable for expensive insurance claims from injured parties. Typical examples of bullying behaviour are these:

a) Your horse is a bully in the field *(below)*, chasing, or kicking out at other horses.

b) Aggressive behaviour in the stable.

c) Your horse knocks or nudges you violently with his head or treads on your feet.

■ Cause

a) An alpha or dominant horse is reinforcing his position in the group.

b) Barn sweet – he feels safe in his stable and does not want anyone to enter this space.

Barn sour – he is trying to get someone's attention as he is unhappy in his confinement.

Above: A horse cannot display natural behaviour when stabled and this confinement can lead to a variety of behavioural problems.

c) Immature, attention-seeking, uncoordinated – see section relating to Clumsy/Boisterous Behaviour below.

We will look at remedies to the three types of behaviour in succession below.

BULLYING BEHAVIOUR IN THE FIELD

■ Remedy

It is helpful to create a pair-bond for an alpha or dominant horse and perhaps just turn out the two together, rather than leave him in a large herd. In this way the horses will rely on one another for security and comfort, rather than being in competition. It is highly likely that one will remain obviously dominant but unlikely that he will ever cause his pair-bond any real harm. Often, in this kind of situation, it is safer and kinder to keep animals of the same sex together to mitigate hormone-fuelled disputes.

Sometimes (as can also be seen with human bullies) such horses are actually bored or insecure in their environment or overworked and stressed and may display aggressive behaviour as a kind of release from these tensions.

Another factor which influences the behaviour of both wild and domestic herds is competition for food. Dominant animals get their fill of food before others, so turning horses out when there is very little grass or hay available will naturally lead to competitiveness.

Splitting The Ration
Either partition off the field with individual paddocks for dominant horses (individually or with their pair-bond) so there is no need for aggressive behaviour or, when feeding hay, make

Below: It is all too easy for horses to be injured when scrapping in the field. A kick (especially from a shod hoof) can have devastating consequences.

Above: Make sure there are more piles of hay than there are numbers of horses to feed and that these piles are spread wide apart.

several more piles than there are horses in the field and place these at least 3m (10ft) apart, preferably further, so it is harder for one horse to intimidate another and drive him from his pile of hay.

Never give individual feeds to a loose group of horses in a field unless you are confident that there will be no squabbling. Either tie up those horses who cause trouble or remove the subordinate ones to their stables where they can eat in peace.

continued ➡

■ BULLYING/AGGRESSIVE BEHAVIOUR

AGGRESSIVE WHILE STABLED

It is horrible and, indeed a liability, to have a horse who lunges teeth-bared over the stable door at you (or at others walking by) rather than being pleased to see you approaching him. Other horses who may be perfect in all other respects can turn their backs to you and threaten to kick, or pin you violently against a wall of the stable. This is frightening for any owner and a liability if you expect anyone else to look after your horse if you cannot make it to the yard for any reason.

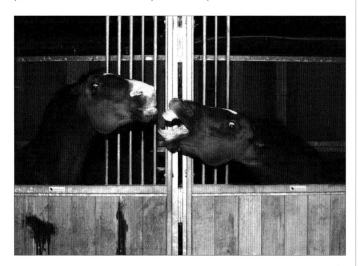

Above: A horse that is aggressive is a liability potentially injuring people and other horses. Do not write him off but try and find the cause of the problem.

■ Cause

Barn Sweet: The horse sees his stable as his 'personal space' and is threatening to those who encroach upon his comfort zone. The 'barn sweet' horse has become introverted and needs to interact with humans and other horses so that he can be stretched both mentally and physically to stimulate enthusiasm and well-being. It's a bit like having a Border Collie who wants to work but which you decide to leave all day in your house with nothing to do. Although it is genuinely pleased to see you

when you do appear, it may snap at you. Should it be labelled vicious or is it merely displaying pent-up frustration?

Barn Sour: The horse is suffering from a lack of stimulation, mentally or physically, and is actually stressed and displaying frustration at his confinement.

Stress: An overworked horse or a horse who is never allowed to 'chill' cannot switch off. Horses cannot appreciate a day off unless they can indulge in their natural behaviour. For this reason giving a horse a 'day off' from work while he remains stabled is not acceptable horse management.

Right: A balanced lifestyle with freedom to 'be a horse' and display natural behaviour as well as a varied training regime and consistent handling will make for a well-adjusted and contented horse.

Above: Aromatherapy oils can be used to relax or stimulate your horse's mood and improve general well-being and to aid bonding.

Above: Treats hidden in the stable or a treat-ball will occupy a horse in natural foraging behaviour. It is better than just giving him 'meals'.

■ Remedy

Spending more time with your horse plus affording him a stimulating lifestyle, which includes a balance of freedom to display natural behaviour plus having a job to do, will lead to a more contented horse. This may mean getting someone else to visit your horse, spending time grooming it, massaging it, stimulating its senses with aromatherapy oils, giving it a treat ball to play with, feeding a variety of forages etc. A day off from work should be made obvious to the horse by turning him out with another horse in a yard or manege if no field is available. (See also Cross Training)

PINNING YOU AGAINST THE STABLE WALL
Case History

A brilliant all-round riding school horse was in such demand from clients that her popularity led to an undesirable change in behaviour. She lunged at people from over the stable door and pinned them against the stable wall if they entered. On one occasion she even charged with teeth bared at the instructor in the middle of the indoor school during a lesson (much to the surprise and embarrassment of the rider).

■ Cause

An overworked, sick or stale horse will soon lose enthusiasm to work. It is essential to watch your horse closely and pick up on any signs of mental or physical problems before they become deep-rooted.

Right: Back off! Firm but fair handling will curb a domineering horse. Mimic biting him, as another horse would, by poking or pinching to prevent him crushing you.

■ Remedy

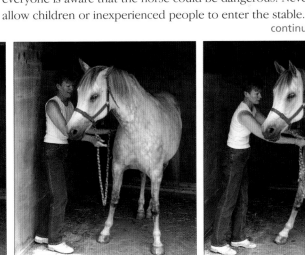

- Try to work out the underlying cause of the horse's unhappiness and take appropriate remedial action.
- When approaching or entering the stable of an aggressive horse, carry a whip pointing ahead of you *(right)* to create a barrier between you and the horse.
- Fit a breast bar across the door opening so you are able to leave the door open while inside and can duck underneath to get out rather than turn your back trying to unbolt the door to get out, which could make you vulnerable to attack.
- Insist your horse moves back when you go inside and is not allowed to mug you for food.
- If the horse invades your space, poke or pinch him to force him to back off. You will be mimicking the action of another horse biting him.

Safety With any forms of aggressive behaviour make sure everyone is aware that the horse could be dangerous. Never allow children or inexperienced people to enter the stable.

continued ➡

■ BULLYING/AGGRESSIVE BEHAVIOUR

CLUMSY/BOISTEROUS BEHAVIOUR

It is not fun to put all your efforts into caring for a horse but to come away from the stable yard every day with yet another bruise.

Case History

I used to have a lovely colt who had a promising career ahead of him. However, on a day-to-day basis he was exhausting. He would be waiting for his feed, head over the stable door and when I approached would somehow whack me with the side of his head, often catching my shoulders or even my face. It could have been very dangerous as he had a lot of weight behind him and I could easily have been knocked out. With annoying frequency he would stand on my feet and he had been known to actually kick out forwards with a foreleg, something that most people would not be prepared for. With a horse such as this you have to have a strong constitution!

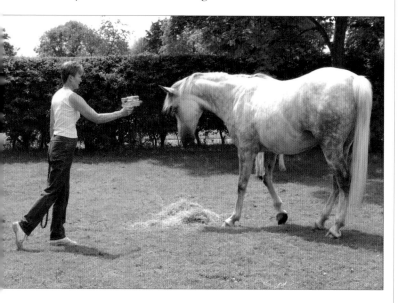

Above: A bad mannered horse is dangerous if he does not have respect for his handler. While I do not advocate violence towards any horse, he must know his place and a swift pinch or a squirt of water will curb pushy behaviour.

■ Cause

It could be (as with my colt) that the animal is immature and simply both unbalanced and exuberant. He is not aggressive but playful and eager to see you and/or the food you bring. A colt, by nature (being an entire male), will have the strength and intent perhaps to knock you to the floor. An older animal may have been allowed to continue this behaviour which is not necessarily aggressive, just overly keen. Coordination is related to fitness, concentration and discipline so a novice or 'soft' horse which has been let down (not being in work) may be clumsy.

■ Remedy

Balance and coordination come with practice – learning commands, sequences and general coordination skills plus body awareness can all help alleviate this condition. These can be taught with groundwork which will lead to a horse that has respect for you as a handler/rider. This, of course, is something that needs to be established from a safety point of view and which will assist in all the training exercises that you undertake together. I have found that the 'Seven Games' of Pat Parelli's Partnership System provide an invaluable grounding for young and boisterous horses. Polework exercises also help improve the co-ordination of clumsy horses (see illustrations).

Above, left to right: A clumsy horse will benefit from coordination exercises which will make him more self-aware. Leading through a grid of poles is an excellent task to keep the horse's attention focused and aware of each footfall. Repeat the exercise in both directions until he can pass through the corridors without stepping outside or knocking a pole.

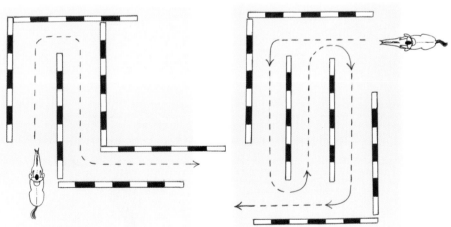

Above: Six poles can be used to make a corridor, each channel being about 1m (3ft) wide. The exercise should be done in both directions, firstly in hand and then ridden.

Above: Progress to more demanding configurations. Tight turns may require halting or backing up to adjust the horse's hind quarters. This pattern of poles can be ridden in many ways.

Above: The width of the corridors can be narrowed as the horse becomes more proficient but make sure you know your own position to avoid being trodden on!

■ AGGRESSIVE TO FEED IN A GROUP

■ Problem

Feeding a group of horses in a barn or field can often be a problem. As an owner it is essential to recognize potentially dangerous behaviour, not only for the welfare of the horses but for your own safety, so that you can avoid being trampled or barged by horses squabbling to get at the food. The most gentle of animals can sometimes act as if crazed in their efforts to mug you for anything you are carrying.

Above: A more dominant horse may well force others away from any food. It is essential not to just put down feed buckets and leave, but to watch that each horse gets its fair share. You may need to tie up troublemakers.

■ Causes

Hungry horses: Obviously if horses are hungry, their survival instincts take over and aggression can be their way of asserting status in the hierarchy of the herd to achieve first pick at the food. Old or less dominant horses may suffer as a consequence.

Bad manners: If you allow horses to invade your space and grab at the food you carry, then bad manners will be learned and the behaviour can become ingrained.

■ Remedy

- Ample forage should always be available so horses can trickle feed (see also the section on Feeding Naturally for the benefits to both health and behaviour of this approach).
- Feed hay before hard feed.
- Put down more piles of hay than there are horses and allow ample room between piles for horses to be able to stand clear of flailing hind legs!

Above: Insist your horse does not mug you for food or titbits. Make him stand back while you put the feed bowl down or carry a whip to keep several horses at a safe distance.

Below: Horses inhabit an imaginary 'personal space'. They only feel comfortable allowing familiar and trusted companions into it, especially when eating or drinking.

Circle about two horse's length in diameter

Behavioural and Handling Problems

- Carry a schooling whip in the field, smack it on the ground and insist that no animal gets too close to you.
- Feed all horses at the same time or section off those who do/don't have the same number of 'meals' a day.
- Feed the most dominant horse first.
- Hand out the other feeds in order of herd rank, well separated from each other.

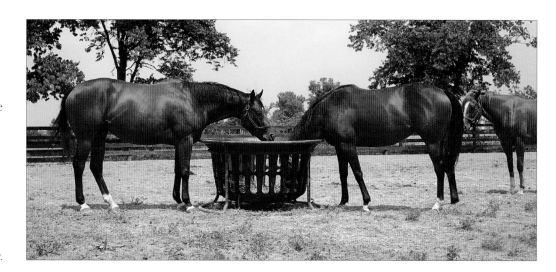

Above: Horses will soon learn their 'positions' and queue for feed or water. They can often be seen to wait in the same locations and a wise owner will observe this ranking and distribute feed in a similar order to avoid quarrels.

- If you feed the horses in the same places for a few days most will learn their 'position' and stand patiently at that site to await their turn.

Tips

Carrots can be fed (even in quite large quantities) to horses and ponies who are not receiving cereals but are overweight or will get upset being missed out at feeding time. Low-calorie short chops can also be given.

A very useful trick is to give dominant horses more bulking food (short chop e.g. chaff/alfalfa) than the others and to mix their cereals and pony nuts well in. With the subservient animals put the chaff at the bottom and the cereals on the top and do not mix. In this way the dominant horse will take longer to eat his food; if he then shoos the weaker ones away from their food, they are more likely to have already eaten the valuable cereals – there will only be chaff left and the bully will return to his own bowl.

- If you have a real troublemaker, tie him up *(above)* or take him out to feed him, only releasing him when the others have finished their food.

■ BAD TO CATCH

■ Problem

You want to give your horse time out in the field but when you come to catch him he gives you the run around. By the time you have finally caught him, you are both sweaty and exhausted and not in the right frame of mind to ride. If you have limited time (and patience) this can be a real problem.

■ Causes

- The horse does not want to leave the rest of the herd in the field.
- The horse is enjoying the fresh grass and being outside and does not want to come in.
- What began as a 'game' with you (which he won) has now become a habit.
- The horse is overworked or dreads being ridden.

■ Remedy

1 Basically, if a horse enjoys being in the field more than being with you, then why should he want to come to you and be taken away from this? Therefore, your aim is to make him want to be with you and want to be ridden. Firstly you need to have a good relationship with your horse. This involves truly understanding him and recognizing when there is something wrong, when he is stressed or is having a bad day and when he wants a scratch or massage to make him feel good. Spend time

Above: For an owner, there is nothing more frustrating than a horse that will not be caught in his field, especially if your time is limited for riding or you are attempting to put the horse away before bad weather arrives.

Right: A horse will easily outpace a human and his charging round is likely to stir up any other horses in the field. Some horses just like to play games and 'protest' for a few minutes, whereas the behaviour can be a real and persistent problem if offering food or leaving the horse out alone overnight still does not encourage him to come to you.

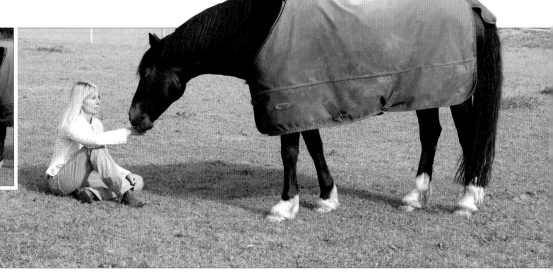

Above and right: Horses are curious and generally crave company. Keep the horse in a field on his own and he will relish your companionship.

Below: Getting one horse out of a herd can be difficult and dangerous if they all get wound up. It helps to spend time within the herd rather than purely turning up to catch your horse. Alternatively, section off part of the field or make a corral where the horse is fed and can be confined when necessary.

with him out in the field so he is not automatically suspicious when you turn up. Picking up droppings or sitting reading or sunbathing in the field will arouse his natural curiosity and he will come up to sniff you. Make the most of this opportunity to stroke him or give him a titbit if he approaches you but do not pester or follow him around.

2 You cannot blame a horse for choosing to stay out in the field where there is good grass, company and freedom – it is quite likely to seem preferable to going into his stable or being ridden. Horses do not have 'ambition' as we do and are quite happy with forage, friends and freedom as the mainstay of their lifestyle.

Rather than turning him out less because you fear you will not catch him, turn him out into a smaller paddock or corral, still with or next to other horses. A larger area can have a temporary fence allowing or preventing passage from the small corral into the main area. The horse can be encouraged (or given his morning feed) always in this sectioned-off area and the gateway to the main area can be closed behind him on days you wish to ride him. This allows him freedom most of the time without the battle of chasing him round a huge field.

continued ➡

■ BAD TO CATCH

3 Once a horse learns that you will either give up in your attempts to catch him if he resists, or get cross when you finally do manage it, this usually reinforces the apprehension that being caught is something to avoid – if they 'win' and you fail to catch them, they get to stay out in the field; if they lose, you

Below: A leather halter can be left on the horse while turned out in the field to make catching easier. Leather will break if the halter becomes entangled.

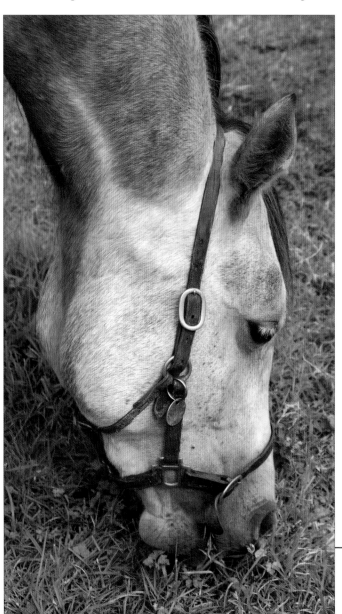

shout at them and tug on the rope. If they are quite happy to come up to you at any time other than when they see the head-collar or bridle then, to make life easier for you, turn them out already wearing a headcollar. Make sure it is a leather or special 'turn-out' headcollar which will break if it gets snagged or caught up. Attach a 30cm (1ft) lead (either cut down a leadrope or use baler twine) and leave him with this on. I am not overly keen on the use of neck straps as it is difficult to judge how tightly they should be secured to be effective but safe.

When you spend time with your horse in the field, aim to play with him and catch and release him a few times. Never snatch at the rope but stroke his neck or head and then pick up the rope, lead him a few strides and let go again.

Be persistent. For a horse that keeps running off, carry a long rope or lungeing whip and insist that he keeps moving. Normally he would run off to a safe distance and then stop and graze. With the rope or long whip you have the advantage that you do not have to get so close. Wave it at him the second he

Above: If you frequently visit your horse just to stroke him or feed him a titbit, he will begin to look forward to you turning up again.

puts his head down to graze and insist he moves off again. Act 'big', wave your arms. Stop only when he approaches or turns to face you; then lower your arms and look downwards. Slowly approach him keeping the submissive stance and stroke him (and catch him) if he allows. If he moves off, insist he keeps moving again until he turns or approaches you. Keep up this nagging, not allowing him to stop and graze. He should learn that it is preferable to be caught than to be persistently chased

Right: Horses respond well to the 'advance and retreat' method. If they run away then keep driving them on. If they turn towards you, encourage them in with arms and eyes lowered.

off. Not being caught is no fun, but he gets a stroke and a titbit if he does agree.

4 I have found from experience that horses that do not want to be caught usually do have a valid reason. One of my horses was an ex-riding school horse – her normal workload was not a pleasurable hour-long hack but a minimum three hours of going round in circles with various bad riders on board! Another horse who developed this behaviour was a young Arab who, having only been recently backed, had weak back muscles and a tender mouth. Being ridden was actually going to be painful until these areas hardened up. She was very cold-backed but enjoyed learning in the arena and appreciated the new sights when out on a ride. It was important not to let this genuine association with pain develop into a bad habit that persisted even when she was fit and felt no discomfort. Accordingly she wore a headcollar in the field and I often went in to play with her or catch her just for a groom (and a tickle on her belly which she loved) or bring her in for a feed.

In essence it is important to look behind what may be causing the vice and seek to adjust your management of the horse and understand and be sympathetic to his signals. **What is the horse trying to tell you?** Horses tend not to expend energy without good reason and therefore there must be a problem if the horse would rather run round the field than be with you.

Right: If you are fortunate enough to find 'that' spot where your horse loves to be scratched, you have a distinct advantage in making friends!

Have his back and the fit of all his tack checked. Get the vet to look at his mouth. Many horses have ulceration in their mouths from sharp teeth that will be aggravated by a bit. This problem is easily remedied by a visit from the vet or equine dentist.

■ BAD TO LEAD – PULLING AND DRAGGING

■ Problem

If your horse is a handful to lead, you will not feel fully in control. This is never a good position to find yourself in as a horse owner. It is a problem which will impinge on all the daily horsecare tasks that you undertake as well as compromising the safety of others.

■ Cause

It does not take a genius to realize that a large animal such as a horse has a distinct advantage of strength over any handler.

Even ponies can run rings round us if they choose to and I have suffered the embarrassment (plus a scar on my hand to prove it) of having been dragged along the floor on my chest by a yearling pony wanting to follow its field companion! Although a lack of respect for the handler is generally the overriding cause, pent-up energy, over-enthusiasm and even fear can play a part. Conversely, a horse or pony who hangs back and needs to be hauled along by the handler is probably displaying a lack of energy or enthusiasm and possibly a sense of dread.

Below left: Mind over muscle! You may want your horse to come in from the field but what do you do if half a tonne of horse has other ideas?

Below right: Too late! Once this lesson is learned by the horse, he will know how easily he can overpower you in a straightforward trial of strength.

Remedy

The wonderful thing about horses is that, for the most part, they want to be guided by others. They are herd animals and gain confidence from being close to other members of the herd, and by following its more dominant members.

Attitude and understanding are the keys. Our physical weaknesses can be overcome by pre-empting how the horse may react and giving him clear and confident direction. Therefore, in my situation when leading the yearling, I should have appreciated that the youngster would want to follow when his companion was put into canter in front of him. I would then have had the foresight to take appropriate action.

continued ➡

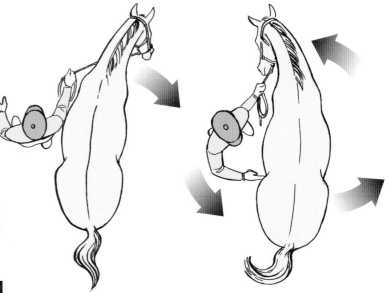

Above left: When a horse turns his head one way, his quarters will move in the opposite direction. Therefore if he turns away, you will lose any leverage and control.

Above right: By attaching a rope to the side nearest you and keeping his head facing towards you, you can regain control and dictate where his hindquarters will go.

Below: Under control. With a chiffney or a pressure halter, the horse cannot pull away so easily and the handler can now safely lead two horses.

■ BAD TO LEAD

Helpful tips to minimize the physical disadvantage we suffer from and to maximize control are:

- Create a good relationship and practise your groundwork skills and play with your horse in the field or manege.
- Use clear vocal and directional commands in hand e.g. 'Back' while pushing his chest.
- Back up these commands if they are ignored. Act big – shout, stamp, wave your hands (or a whip to add distance) – do not let him ignore you.

Above: A chiffney is a type of bit attached to the bridle commonly used for leading horses, loading horses into trailers or handling stallions.

Above: Follow your nose. Passing the lead rope around the nose aids leverage and control.

- Always wear gloves if there is the slightest chance that the horse may run off. Rope burns are extremely painful.
- Keep a contact but do not hang on the horse or he will pull against you.
- Clip the lead rope to the side (rather than the back) of the headcollar. Alternatively, bend his head slightly towards you and push your elbow into the side of his neck. This gives you lateral leverage if the horse pulls away and helps to prevent him dragging you.

- Have you tried 'controller' headcollars? They exert pressure on the horse when he pulls, and release when he does, and therefore he fights himself not you.
- Leading in a bridle gives more control. A chiffney bit (a large looped bit that acts on the tongue) should only be used in experienced hands or it could lead to the horse sustaining a broken tooth or even a jaw fracture.

Safety Advice: Accept the fact you may be 'over-horsed'. It can be dangerous for a novice to take on an intelligent Alpha horse, especially if the handler/rider lacks the confidence and strength to control him effectively.

Above: A chiffney should be used sensitively so the bit only comes into play when the horse itself causes it to do so (e.g. by rearing or pulling).

■ STUBBORN AND SLOW

■ Problem

Rather than guiding the horse where you want him, it feels like you are physically hauling him across the ground! These horses appear slow and uncooperative, making life exhausting for their owners

Above: An unwilling horse is often unfit or lacks motivation. Try to identify what is causing the resistance and whether it is laziness or genuine pain.

■ Cause

The three main causes are lack of motivation, fear and actual pain. Ask yourself why your horse does not want to be led. Could it be he does not want to leave his friends in the field to do some schooling? Is he bored with being fussed over (some horses actually do not like being groomed)? You may want to load him into a trailer, or tie him up next to your friend's horse, or give him a bath – but does he? He may not be uncooperative but actually suffering physical discomfort.

Left: A well-mannered horse will not cause a problem if you have to ask other people to take care of him during your absence.

Above: A bad experience – for example with water or a hose – can remain in the memory of a horse and trigger seemingly irrational fears.

■ Remedy

Know your horse. Recognize visual signs of discomfort. Perhaps the yard is cobbled and he finds the surface uncomfortable to walk on or he is stiff in his joints with arthritis. It is not helpful just to drag him along. Recognize that he may take a few minutes to loosen up and make allowances for that.

Lack of motivation could stem from fear. You may want to tie him up next to your best friend's horse but they may not have a similar fondness for one another. Perhaps your horse is bullied and will feel very stressed in close proximity to a dominant horse. You may get cross with him if he will not go near the hose which he has seen a million times, but a cold water bath does aggravate rheumatism and he may be shying away from this memory. He may perform well but does he want to perform for you?

Once you begin to see life from his point of view, you will have made a huge step forward. Your consideration will enable your horse to enjoy life more and be motivated to be with you.

■ BAD TO LOAD AND TRAVEL

■ Problem

A horse that will not load into a lorry or trailer puts severe constraints on the activities that you can enjoy together. Other than very local events, it is usually necessary to travel to take part in all types of competitions or excursions. You may also need the horse to travel to an equine hospital for treatment. Another huge drawback is that it will affect his resale value if the new owner cannot even take him away in a trailer, let alone transport him to the events they aspire to attending!

understandably reluctant to enter it a second time. It is such a common problem and, having watched people struggling to overcome it for up to four hours after a show, I feel for both horse and owner going through that experience.

■ Remedy

Trauma from past experience cannot be eradicated, but you can generate renewed trust to help the horse to re-associate loading

Above: Even a pony can be a handful if it is reluctant to load. Being put into a confined space is not an inviting prospect, especially if friends are whinnying in the background. The handler should always wear gloves to avoid rope burns.

■ Cause

Horses have good memories and will not put themselves willingly into a situation that they perceive as frightening. It takes a calm, trusting horse to walk willingly up an unstable ramp into a dark, confined space and then have to balance while being buffeted around during the journey, finally to be unloaded into a strange environment away from the sights and smells he knows. Added to this, many horses have scraped a leg falling off the ramp or been pulled in the mouth or whipped in an effort to get them into the 'nasty box' and so are

and travelling with a more positive experience. Problems are generally caused by people not addressing the problem until they **have** to put the horse in a lorry or trailer. Two hours before a show is **not** the time to start training the horse to load.

Above: Regularly feeding him in the trailer or lorry will teach him that it is a pleasant area and he will soon be marching in without a struggle.

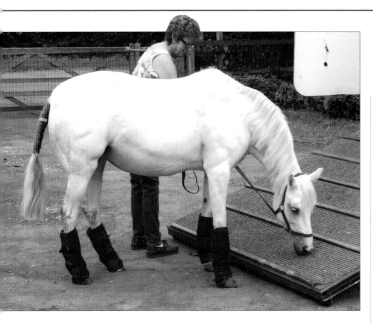

Above: Allow the horse time to sniff the ramp. The more solid the ramp is the better, as a slippery or unstable ramp which rattles will not boost confidence.

Using a whip, broom, lungelines or the help of ten assistants may be a quick fix, but it will only serve to reinforce the negative associations in the mind of the horse.

Be calm and positive yourself

If you are in a rush and nervous before a competition, then your horse will pick up on these anxieties. Raised voices and general fussing will heighten tension. If necessary get someone else to load the horse.

Make loading a 'normal' activity

One of the commonly used obstacles in the PTV (Parcours en Terrain Varié) section of Le Trec is leading in and out of a trailer and so it makes sense to practise with your horse at home. If possible, remove any partitions and regularly leave the trailer/lorry in the field or turnout yard (N.B. most trailers would need to be attached to a car with the jockey wheel and axle stands lowered for added security) Horses are inquisitive and, given time, will sniff and explore the vehicle, especially if there is a straw bed or some food left inside. They can even been given their feed one at a time inside (start with horses

known to load well) and then led out of the front ramp while the next horse is encouraged to go in. Practise several times leading the horse in, standing for a few minutes with some feed and then out again. Do not close up the back or move the lorry/trailer until he is happy with this.

Security and companionship

If possible, take an equine 'friend' along for the ride. It will settle jangled nerves. I stress that it must be a horse or pony that they get along with. I once felt terrible because I travelled a mare with another strange horse and when we unloaded her all her whiskers had been chewed off and she had bite marks on her face – a bad experience that would not have helped in the future.

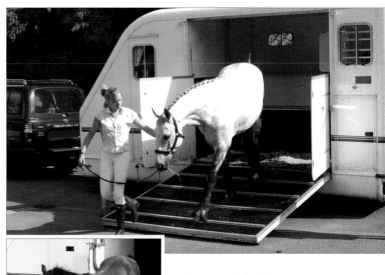

Above and left: Take a companion along for the ride. This dressage horse benefits from the calming influence of a retired pony friend who, in turn, gets a renewed purpose in life, a change of scenery and the enjoyment of the 'buzz' of the event.

continued ➡

■ BAD TO LOAD AND TRAVEL

Above: Horses are calmer when travelling in a group. They create a lot of heat so ensure ventilation is good but only open windows fully when stationary.

Comfort and safety

Provide a haynet full of good hay – not the haynet left in the trailer from the last journey that is dusty and mouldy. Rubber mats on the floor *(above)* or a bed of shavings or straw will aid grip and balance and help to cushion any jarring from the

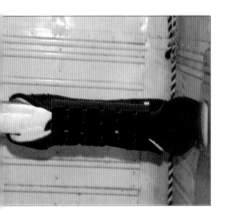

Above left: When you brake, a horse can be flung against the breast bar which is often an unforgiving metal pole. Create greater comfort by wrapping spare leg wraps around the bar. They are perfect for the job.

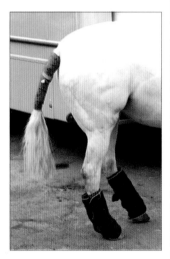

Above right: If possible, use leg and tail bandages to protect the horse from scrapes or rubbing.

movement of the vehicle. Check for protruding screws which frequently rattle loose on older trailers and can injure the horse. A horse will be thrown forwards, backwards and sideways onto the partitions and breast bar by the vehicle's movement so limit bruising to the horse by making sure that these are well padded. Foam or inflated partition cushions make the journey more comfortable and I find that wrapping spare travel boots round the breast bar helps to avoid the horse getting bruised from a sudden jerk.

On the subject of travel boots, it is always advisable to use these plus a tail bandage on the majority of horses. Travel boots or bandages will protect against the horse grazing himself or other horses travelling together from being scraped by a hoof, as they try to brace themselves to maintain their balance. Horses often lean on their rumps to avoid leg fatigue so the dock and tail hairs need to be protected from any rubbing. A poll guard will protect the top of the horse's head, especially important if you have a tall animal or one with a tendency to fidget or rear up.

I should stress, however, that attempting to put leg boots or bandages on an animal that will find this a stressful experience is far more likely to cause problems than solve them. These may increase the horse's feeling of claustrophobia and actually cause an accident if the bandage wriggles loose as he frantically stamps or shakes his legs. It is best to try loading the animal with as little fuss as possible. Practise putting boots on him another day when he is calm after his regular grooming session.

Bear in mind the ventilation and temperature conditions when deciding whether to use a rug while travelling and remember that several animals travelling together are likely to generate heat, especially if they are tense or sweating from exercise. A sweat rug or fleece cooler is normally all that is required.

Left and below: It is easy to get injured by a horse rushing or leaping out of the box. Rather than having to let go of the rope or getting dragged, use a lunge line rather than a lead rope so there is plenty of slack to let out.

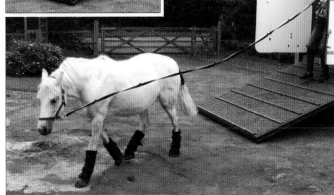

Left: Do not expect your horse to wear his travel boots for the first time on the morning of the show. Practise putting these on him on a day when you are both calm and there is no urgency to get the job finished to a particular timetable.

Remember

- Park the vehicle in a gateway or block easy escape routes on either side of the ramp before starting to load.
- Open up the front doors/ramps and move partitions across to make the interior look less dark and claustrophobic.
- Have a haynet in the box ready and pony nuts for rewarding any steps up the ramp.
- Think positive, this time he may go in first time!
- Always stand at the side of the horse and not in front of him. He needs to see where he is going.
- It makes sense to wear gloves, a hard hat and tough footwear in case he leaps about.
- Tie him up short enough not to be able to squabble with other horses but loose enough so he can move his head.

Tie him to a loop of baler twine which will break if he should unbalance and fall over.

- When you unload him, allow him his head as he may want to examine the ramp or step down. Be prepared that he may leap off. I have found that using a lunge line is preferable to a short rope for unloading as you can still keep a hold but let out more rope if the horse leaps to avoid being dragged behind or crushed by his weight.
- Make short trips first of all and ideally take his favourite companion. Make it a pleasant outing – travel to a local forest or beach for a change of scenery.
- When you unload, allow him to graze or offer a haynet while you tack up. Always carry a container of water with you and offer it frequently as he is likely to lose fluids through exertion and anxiety.

■ BAD TO CLIP

■ Problem

Many horses in medium to hard work benefit from having some of their thick winter coat removed to prevent excessive sweating each time they work. It is also impossible to groom a horse covered in wet mud and the prolonged length of drying time for a horse with a long coat that is saturated in sweat or rain/mud can make them susceptible to a chill.

Regrettably you are likely come across talented horses who have changed hands several times when their new owners have discovered that they will not stand to be clipped and are therefore at a disadvantage when in hard work.

Another factor to consider is that the use of electric clippers can be dangerous because they involve fast-moving blades and dangling electric cables so it is essential that a horse cooperates when being clipped for the safety of all involved.

■ Cause

It is understandable that the buzzing/rattling sound that clippers make will be alarming to horses, possibly sounding like some giant predatory insect wanting to suck their blood! Any horse that has previously had a bad experience may well have been nicked by the blades or scalded when the clippers were allowed to run too hot. Standing immobile for the length of time it can take to complete a clip (40 minutes to two hours) is a challenge in itself for many highly strung animals.

■ Remedy

Before attempting clipping for the first time or if you are dealing with an animal known to resent clipping, you will need to desensitize the horse to the scary noise that triggers the fear. Clip stoic animals close to your horse's stable so he recognizes it as a habitual sound in the yard. Get someone to run the clippers firstly outside and then inside the stable when you take his food in (but do not prolong it while he is actually eating if he is naturally defensive with his food or he will view the noise as an annoyance).

Progress to holding the clippers close to the brush when you groom him so he gets used to the sound in conjunction with something passing over his body.

Above: A nervous horse can be a danger when it is clipped, especially where there are electric cables and sharp blades involved.

Right: Allow the horse to investigate the clippers while they are off.
Far right: Run the clippers next to a grooming brush so he gets used to the noise and something moving over his body before cutting any hair.

TOOLS – TOLERANCE – TECHNIQUE

Tools: Before you begin you should have the following in place:

- A clean, dry, groomed horse (any dampness or sweat will cause the blades to snag and pull at the hair).
- A serviced set of quiet clippers, safety checked to ensure the cable (if electric) has no nicks or kinks that could cause shocks or power failure and, in the case of battery-powered clippers, that the battery is fully charged.
- A newly sharpened set of blades plus a spare set.
- Clipping oil.

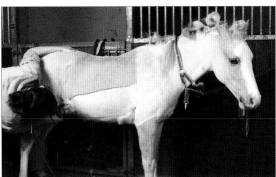

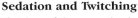

Both above: Begin at the neck or shoulder and be considerate when working on the belly and other ticklish areas or where there are folds of skin.

- A bright stable (remove any free-standing water buckets), or a secure yard area.
- A radio to distract the horse and mask the noise of the clippers.

Tolerance: Handler and horse should be in the best frame of mind therefore:

- Do not begin clipping unless you know you have enough time to complete the job without rushing.
- Provide a bucket of chaff or pile of hay in a wall-mounted feed bucket to occupy the horse. I prefer this to hay in a net as the horse does not need to jerk the forage out of the net and therefore remains stiller.
- For your first time just do a 'bib' or neck and belly clip (like a low trace) and do not try to take any hair off the face. The less complicated the clip, the shorter the time the horse has to tolerate the clippers.

Technique: Get an experienced person to wield the clippers so you can learn from their skill.

- Adjust the tension of the clipper blades, oil regularly and brush loose hair away. Switch off the clippers if the blades are running too hot and allow them to cool before resuming.
- Begin in the neck area and be considerate when working on the belly and other ticklish areas.
- Ask an assistant to pull forward each foreleg when clipping the elbows, or you are quite likely to nick the skin. The pasterns and heels are other areas which should be approached with **extreme** care.
- If things are not going well, it may be better to call it a day and finish the clip later.

Sedation and Twitching

If either of these methods are to be employed, it is essential that (oral or intravenous) sedatives or a lip twitch are applied to the horse **before** he becomes agitated (see also the section on Sedation on pages 96-97). People attempting to placate an already stressed horse put themselves in grave danger as many horses fight desperately against such measures and it only serves to dramatically worsen future behaviour.

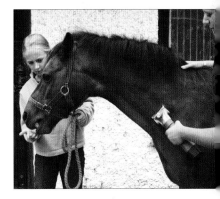

Above: Ask an assistant to feed treats to the horse to keep it calm and reinforce the idea that clipping can be pleasurable.

■ BAD TO CLIP – MANE & TAIL PULLING PROBLEMS

■ Problem

For many show classes and competitions it is a requirement or 'the done thing' to present your horse with plaited mane and pulled (or plaited) tail. A plaited mane crowns the neck to give a finished, tidy appearance and pulling or plaiting the tail tapers the unruly hair at the top of the dock and accentuates the hind quarters. There are many breeds including Arabs, Native ponies,

Above: I am always impressed by a beautifully plaited tail that shows the skill of the groom and which enhances the appearance of the quarters of the horse.

Friesians and Standardbreds where the natural appearance is embraced but most owners have cross-bred animals who, not fitting within one of the breed exemptions, are expected to conform by presenting their horse plaited for competition.

In order to plait a mane it cannot be too long or too thick or each plait will look like a tennis ball! Therefore hairs have to be removed from the underside of the mane to shorten it to the desired thickness and length (usually about 10cm/4in). There is no doubt that pulling the hair like this can cause discomfort to the horse

who can soon learn tricks like throwing his head around or crushing you against the wall of the stable to prevent you from getting hold of more hairs.

For a plaited tail, the long hairs on either side of the dock are used but unfortunately good plaiting is an art and cannot really be done in advance as the plait can be ruined by the horse rubbing it or the tail guard slipping down during transport in a trailer. A 'pulled' tail gives a tidy appearance that lasts far longer and most professionals favour this method. Consequently they are copied by amateurs, but regrettably not all horses appreciate the titivation!

■ Cause

We all know that there is a reason why horses have long manes and long tails with hair fanning out at the dock. In summer the hair can be swished to dislodge flies and in winter the hair helps to dissipate rain as well as keeping these areas warm. You will commonly see horses with their bums to the wind in foul weather and a horse with a pulled tail is at a distinct disadvantage in such situations! Having said this, it is unlikely that horses have the reasoning power to understand the consequences of the removal of mane and tail hair. Their only concern (which is likely to stick in their memory) is the pain caused by having hairs tugged out by the roots – something surely we can appreciate as being a highly unpleasant experience.

■ Remedy

In order to minimize the horse's adverse reaction to hair pulling, we basically need to lessen the pain he experiences. This means that you may well have to do the job in several sessions, pulling out only a few hairs each time and **always, always** in warm weather or when the horse is warm through working and his pores are open.

If your horse has over-wintered with a long mane and now you want to tidy him up, it will cause great distress if you attempt to reduce the mane from around 23cm to 10cm (9in to 4in) all in one go as so much hair would have to be removed. I have had surprising success by firstly cutting the mane (yes, with scissors!) to pretty much the right length and then pulling (or thinning) the remaining hair to soften the line and achieve the 'pulled' appearance. It must be remembered that the success of any makeover is directly proportional to the skill of the 'expert'!

Right: You can pull small sections of mane while mounted. Do a little after work when the horse is warm and the pores are open.

Behavioural and Handling Problems

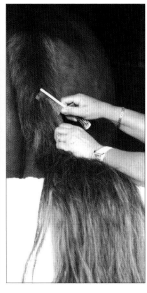

Above: Stand out from the crowd! A running plait is a stunning alternative without having to butcher a long flowing mane.

Safety First

To break the cycle of the horse crushing you against the wall, try pulling a few hairs while riding him but remember to wear gloves or you too will be in pain! When pulling a tail, use a barrier such as straw bales *(above left)* between you and the horse or, alternatively, back him up to the stable door and work on the tail from outside to prevent him from kicking you *(above right)*. To prevent possible injury to the horse if he leans back or kicks the door, use travel bandages covering his hind legs. At all times have an assistant who can hold and reassure the horse and offer titbits as a reward.

Tip

A local anaesthetic cream can be purchased from a pharmacy. This will numb the areas making it a far less painful experience.

Below and right: With such effective thinning tools, there is no reason to cause pain by pulling. Bladed thinning combs created excellent results if used skilfully.

Totally Painless

I am a fan of pulling/thinning combs and hand trimmers and when used cleverly the effect is indistinguishable

from the hair being pulled. Safety razors and thinning blades are superb for removing stray hairs and can be part of the weekly grooming routine to remove growth and keep the effect looking good.

Stunning Alternatives

If you would like your horse to look tidy but are not so concerned by what is traditionally done, 'running plaits' plus ambitious-looking variations are becoming more popular. These preserve even the longest, thickest mane with its natural benefits but make a well turned-out style achievable for those occasions when you want to look smart.

HEAD SHY AND EAR SHY

Problem

A horse or pony that flinches every time you raise your hand near its face is going to be a problem in terms of putting on a halter or bridle or applying fly repellent in this region. It can become such a problem that horses have been known to rear up to get away from people who are trying to touch their face or ears and while remaining calm in other respects.

A bad experience may cause a horse to mistrust humans who try to touch his head.

Ears: *Can become sore from rubbing by an overly tight headpiece or browband. Headshaking and ear rubbing can be symptoms of flies or parasitic lice and mites. A vet should examine if any of these symptoms persist or if excessive brown earwax is present.*

Eyes: *Can be irritated by hair from a long forelock, grass seeds and flies. If following behind another horse, it is common for dust to be kicked up into the horse's face by the horse in front. Horses with red or continuously watering eyes should be examined by a vet as they could be symptoms of conjunctivitis plus those who have been seen to knock into things or have problems jumping may have cataracts (common in older horses).*

Cheeks: *The thin skin over the bone of the cheeks can easily be rubbed by the buckles of bridles and halters.*

Nose/Muzzle: *Horses with white faces or blazes usually have pink skin and on the nose and muzzle this is particularly susceptible to sunburn causing soreness and scabs. Inhaled pollen and dust can cause irritation and breathing problems.*

Jaw: *Ill-fitting nosebands and curb chains can cause soreness plus tension in the jaw.*

Mouth: *The corners of the mouth can suffer cracking and splits from abrasion of bits and flash nosebands. Sharp or cracked teeth will cause mouth ulcers and impede the mastication of food and should be checked annually by a vet or equine dentist.*

Below: *This horse is chewing the rope which shows that it is worried about having its head touched. Tolerating this behaviour will give the horse some comfort.*

Behavioural and Handling Problems

Causes

Previous Pain: Perhaps some tack has rubbed and caused a sore or the horse has got his halter or bridle hooked up on something in the past which has bruised or really scared him. His mouth or jaw could be sensitive through dental problems and that anxiety about pain is making him reluctant to be touched. Bad Treatment: Possibly the horse has been treated badly in the past and may even have been beaten. In this case a raised hand, especially near the head, is bound to make the horse fearful. Always approach a horse with arms lowered below muzzle level.

Remedy

Remember that you probably do not know everything about your horse's past. Be sympathetic and try to help him associate touching his face with a pleasant experience, a treat for example. Rub his neck with a cloth or your hand and over time work up towards his head. A training halter may be helpful to teach him that if he lowers his head and allows you to stroke him between the eyes, then he can be fed a treat, rather than throwing his head up where he will feel the pressure from the halter.

Check all tack and make sure not only that it fits, but that it is clean without lumps of grease or mud adhering to it that could chafe. A simple bridle consisting of headpiece and bit is all that is required (the noseband is mainly for appearance). If he has possible dental problems, seek advice from your vet or dentist. A disfigurement or dental problem may suggest that a bitless bridle will be helpful in keeping the tack away from a sensitive area.

Above: Horses seem either to adore or to detest having their ears touched. Try massaging round the base and then working your way upwards towards the tip.

Vision

Unfortunately, our horses cannot speak to us and tell us what is wrong and it may well be that a medical problem such as deteriorating eyesight is causing the horse to become alarmed when you raise your arm near his head. His capacity to focus may be impaired or be failing and this can make him fearful.

Ear Mites

Some horses just simply hate their ears being touched, whereas others really adore it – it's a bit like our own tickle hotspots. Horses have difficulty scratching the inside of their ears but I have certainly seen horses perform the most delicate of operations using the corner of a fence post or tree branch to scratch an itch in their ear. Some show real gratitude when they have their ears rubbed both on the outside and inside. What is often causing the itch is an irritant which is commonly found to be mites or small flies living in the ear canals. These mites and flies suck blood and cause great upset and irritation and should be treated by applying surgical spirit to a piece of cotton wool and using it to wipe the insects out. You do not want the insects to work their way deeper inside the ear canal so make sure that you gently plug the ear with some more cotton wool as you wipe the mites out.

Above: There may be times when you need to touch the ears for medical reasons. Wiping the ears with surgical spirit on cotton wool will remove mites and tiny flies.

■ NERVOUS/HIGHLY STRUNG

■ Problem

This is a very common problem affecting horses and ponies, often those containing thoroughbred or Arab blood. For this reason a child's show pony could display highly spirited behaviour (so making it unsuitable for a child) because it is of Welsh x Anglo Arab descent in order to ensure superb conformation. A horse will not be a relaxing ride if he spooks and jogs at the slightest upset. He is also likely to lose condition if easily upset, a state characteristically indicated by runny droppings and excessive sweating. These horses often display agitated behaviour in their stable e.g. box walking or calling if neighbouring animals are taken away.

Above: If your horse plays up, try to think what factors may contribute to this problem. Is there a reason for his resistance or over-exuberance?

Above: Some breeds, such as the Arab, are well known for their flighty behaviour. This should be channelled usefully into the horse's chosen career rather than necessarily being seen as a negative characteristic.

■ Causes

Although genetics predispose an animal to be more or less highly strung or 'on its toes', physical and mental well-being plus chemical and hormonal changes also play a large part.

Lack of education and socialization from an early age will keep the 'fight or flight' self-preservation instinct turned up to full volume, whereas the sense of fear and anxiety would have been diminished if the horse had experienced more stimuli calmly (i.e. plastic bags, crowds of people etc.) as part of a controlled education earlier on. The tension of a nervous rider or handler will also translate as heightened adrenalin to a horse and amplify spooky behaviour.

Inappropriate feeding is another common cause. A 'scoop of this' and a 'scoop of that' can have enormously variable effects on each animal. I have known a horse get into a feed room and eat an entire sack of mix will no apparent ill effects, whereas others will go 'ping!' on a ration of a couple of handfuls.

Above: A high-fibre diet is best for those horses that are adversely affected by grains and cereals.

Feed must be balanced not only to reflect the work done by the horse but also its temperament. Consideration must also be given to maintenance of balanced body functioning, i.e. maintaining adequate levels of protein, amino acids or minerals lost through sweating. However, the mainstay of the diet must always be – **forage**. This forage should make up a **minimum** of 60 per cent of the diet, and in most instances could make up 95-100 per cent. We are fortunate today that the range of forages available to us – from straw to alfalfa hay or chopped dried grass – covers such a large spectrum of dietary elements that the necessary fibre, energy, protein and mineral levels can be achieved with either none or very few supplements added.

■ Remedy

- Calm but assertive handling will assist in gaining the horse's confidence and early (or corrective in later life) familiarization with scary situations in a controlled environment will make a horse less fearful and reactive.

Above: A soothing voice and stroke on the neck will go a long way to creating the best chance of transmitting confidence and calm to your horse. Giving clear concise aids and being confident yourself will assist the horse.

- These animals are often very receptive to their handlers' and companions' behaviour so a stoic pair-bond *(above)* and purposeful (not fussing), matter-of-fact style of handling will give them confidence. Never pussy-foot around a jittery horse or he will think that there is a good reason to be worried.
- Ensure that you provide a diet high in fibre.
- Oil (e.g. corn, sunflower etc.) can be added to boost calories and provide slow-release energy without the fizzy element that grains bring. Oil is also high in Vitamin E which has been shown to have a calming effect.
- Supplements containing minerals and salts (e.g. electrolytes) should be used in hot weather or if the horse is sweating excessively after hard work to replenish fluid and mineral levels. The addition of Vitamin B1 (thiamine) and magnesium supplements often help to calm nervy behaviour and horses who previously were 'hyper' in the show ring have been seen to settle down after taking these supplements.
- Calming or hormonal-balancing natural herbs may assist in settling behaviour caused by fluctuations or deficiencies.

When all is said and done, however, we must accept that there are some horses that will always maintain their explosive or unpredictable characteristics despite all our efforts to minimize them, and these mounts should only be taken on by experienced riders sensitive to their heightened reactions.

■ FEARS: TERRIFIED OF PEOPLE

■ Problem

It is easy to assume that if you feed and care for a horse it will automatically love you in return but sadly there are horses and ponies that simply do not trust humans and it can be a long, uphill battle to turn them around. They hide in the back of stables *(below)* or become impossible to catch and appear obviously distressed (sweating, shaking, rearing) when handled.

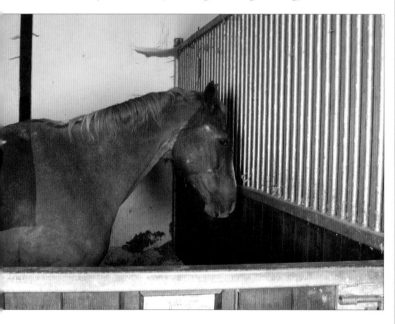

■ Causes

- The animal is young and not used to being handled and feels insecure without its dam.
- Animals can be 'owned' by someone but live and breed in a wild environment (e.g. Dartmoor ponies/Mustangs) and from time to time are rounded up for sale. This sudden switch from a wild to a domestic environment is so alien to them that they feel vulnerable without the safety of the herd and their natural surroundings.
- Previous pain (direct or indirect) is associated with humans or the animal has actually been mistreated. Horses are by nature 'prey' animals who will avoid threatening situations

and, although intelligent, it takes time to overcome this instinct and persuade them to trust the unknown.

As an illustration of how horses can react fearfully to poor handling, I remember going to a dealer's yard where the top stable doors were shut and some of the stables had no windows at all! This was, I was told, to prevent the horses from jumping out. When you entered the stable each horse would retreat to the back wall of the stable, frozen rigid with its eyes rolling – partly from the shock of the bright light and partly because of a general feeling of fear. Some standard-sized stables had three ponies crammed in one stable in the dark but surprisingly these animals were the ones that seemed most sane – probably as they had each other from which to draw security. I am not saying, by any means, that mistreatment is at all common, but what I saw was totally unacceptable and those horses' trauma will stay with them for a very long time.

■ Remedy

Anyone who takes on a young, unsocialized horse or one with a dubious background must have endless patience and try to empathize with the animal. It is not just a case of being the provider of food; we also need to bestow the feeling of safety in order to gain the trust of the horse.

Use a small paddock (or a stable plus a sectioned-off area of the yard), ideally in close proximity to your house or quiet livery stables. It is not advisable to flood him

Right: A horse may have come straight from a hillside and been placed in a domestic environment. Put him somewhere where he can take in his surroundings.

Behavioural and Handling Problems

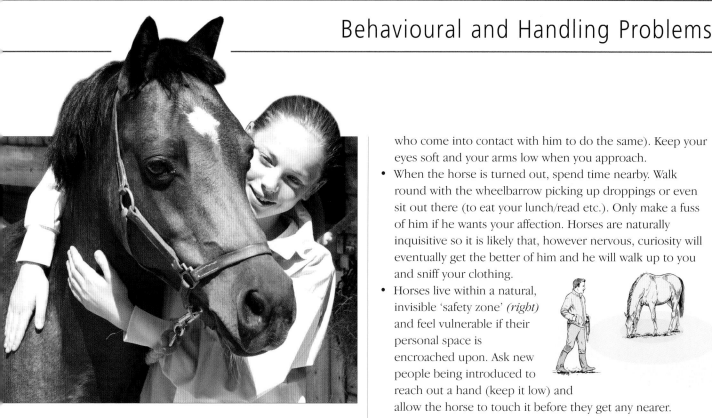

Right: Kindness and consistency will reap rewards. This pony has established that people are not to be feared and has a good relationship with his owner.

with lots of attention but rather allow him to take in his surroundings in his own time, watching people going about their business – horses need to learn about humans too.

Your level of interaction with him should be gradually increased. He should still be kept close to other horses, but initially the time spent with you should be greater than time spent socializing with other animals. If not, you risk having a horse that would rather stay with his friends and who does not want to be caught from the field. You want him to view you as his partner and ally before his circle of companions is expanded. Introduce other people as well by allowing them to take in his feed or stroke him but always remember the following points:

- Keep such sessions short and gradually build them up. Basic things like being touched may be very stressful for him, so begin where he is comfortable and repeat the lessons, gradually seeking to extend his tolerance step by step.
- Use non-threatening body language (and advise other people who come into contact with him to do the same). Keep your eyes soft and your arms low when you approach.
- When the horse is turned out, spend time nearby. Walk round with the wheelbarrow picking up droppings or even sit out there (to eat your lunch/read etc.). Only make a fuss of him if he wants your affection. Horses are naturally inquisitive so it is likely that, however nervous, curiosity will eventually get the better of him and he will walk up to you and sniff your clothing.
- Horses live within a natural, invisible 'safety zone' *(right)* and feel vulnerable if their personal space is encroached upon. Ask new people being introduced to reach out a hand (keep it low) and allow the horse to touch it before they get any nearer.
- Leave (safe) objects such as grooming brushes/rugs etc. in the stable or paddock so the horse has ample time to investigate them in his own time. Don't attempt to groom or rug him for the first time before you've gone through this process of acclimatization – otherwise the sight of the 'scary' object will simply add to his stress.
- He will feel more confident when experiencing new people or objects if a horse is tied up close by. He will gain the reassurance that he has nothing to fear from the calming influence of the other horse.

There are horses who are mentally disturbed and something in their past or mental make-up may prevent them from ever being fully capable of a 'normal' life. With such a horse or pony it may be safer for all involved to accept that the animal is unlikely ever to acclimatize and investigate the possibility of rehoming the animal with a Rescue Charity. The value of these places cannot be praised too highly, both for their dedication to the welfare of animals who would otherwise face a life of misery and for their ability to spend the time with those horses that do need the extra effort to put them on the right path.

■ SCARED OF THE VET, FARRIER OR MEN IN GENERAL

■ Problem

Horses need to see the vet from time to time. Even if they are generally healthy, they will still need vaccinations, teeth rasping and to be checked over when being 'vetted' for sale or purchase. A horse that won't stand still or who spins round his stable is a liability at the best of times and if the vet has to examine him or manipulate limbs, good behaviour is essential.

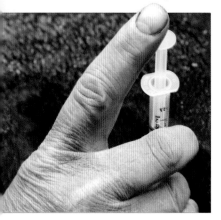

In some cases a horse may be owned and ridden solely by women; and if the vet used is a man, he may start to associate all men as 'vets' and become fearful or aggressive in their presence.

Left: It is easy to understand horses developing a fear of vets if their visits are generally associated with pain. Unfortunately this can escalate to fear of all strangers.

■ Cause

It's a bit like us having to visit the dentist! That smell, the equipment and the dentist himself all add up to a scary experience and if we did not **know** it was for our eventual benefit, we would never by choice sit in a dentist's chair again.

Similarly horses usually only get to see the vet when they are either in pain or, at minimum, to be pushed, prodded and contorted or have a needle stuck in them. This is not a good foundation for a good relationship so it is easy to understand why it is a common problem. Added to this, in the case of illness or injury, the horse is likely to see his owner and other people around him in an upset or fearful state and this fear may then be communicated to the horse.

■ Remedy

We must try to dull the horse's association of these sights and smells to prevent them automatically triggering a fearful response.

Above: A busy yard is a good environment for socializing horses with people. Ask male friends to play at being 'vet' and get them to handle and manipulate the horse while there is no associated injury or pain to create a bad memory.

It is always a good idea to socialize any animal by introducing them to lots of different people, male and female, and this education should start early and continue in every home that a horse enters. Brothers, sisters, mothers, fathers, uncles, aunts and screaming babies should all be introduced to an unsocialized horse!

It is always essential to have a comprehensive first aid kit on the yard. Practise bandaging your own horse and use aromatherapy and coat sprays (as well as some less pleasant-smelling lotions, such as surgical spirit and iodine or 'purple' sprays) to acclimatize him to these when he is well. Progress to asking a friend to manipulate his legs, pinch his neck and pretend to 'treat' the horse and then make a fuss of him and give him carrots.

Vets rarely have time to make free 'friendly' visits but if a vet or farrier is visiting another horse on the yard, ask the owner of that horse and the vet/farrier if he could (perhaps for a small contribution) spend five minutes after treating that horse to say hello to yours, armed with a stethoscope and hoof tester or, in the case of the farrier, pretending to hammer in nails. It is also

Above: The 'clang, clang' of hammering red hot iron and the smell and smoke of burned hoof can be very alarming. Having one leg restrained will understandably add to the feeling of vulnerability. Get the horse used to one element at a time, perhaps trimming, then cold shoeing to begin with.

Below: Begin his education by standing him near a horse being shod so he can experience the associated sounds and smells.

Below: Get in the habit of using strange-smelling sprays while grooming and practise bandaging legs so it will not cause a problem should you have to dress a wound.

effective to tie up the nervous horse in close proximity to one being shod so he can inhale the smoke that is associated with the farrier's work and accustom him to what would otherwise be a very alarming first experience.

This education should be repeated on a regular basis so that antiseptic smells or the tools of the vet or farrier are not seen as something out of the ordinary nor something to be feared. With the irrational fear removed it is more likely the horse will be able to recognize, when the time comes for actual treatment, that these people are here to help him, or at least behave so the vet or farrier can perform their jobs safely.

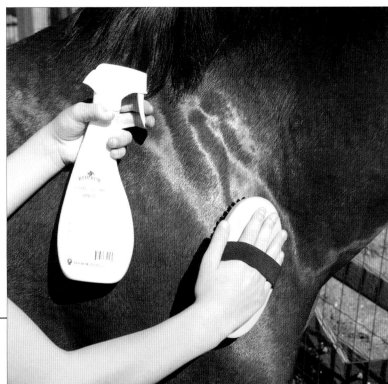

■ SCARED OF/AGGRESSIVE TO LIVESTOCK OR DOGS

■ Problem

It is likely that, at one time or another, we will encounter dogs or livestock when out on a ride but unfortunately these meetings can have a terrifying effect on some horses who will spin round trying to escape or bolt in terror. In other instances horses may actually rear up or attack animals ranging from dogs to donkeys.

Below left to right: It is a lot to ask of your horse to pass through a herd of cattle and only natural that this horse is none too confident. Only a very trusting horse would comply with the rider's request to approach the cattle. This rider could have dismounted and attempted to lead the horse through but thankfully she has taken the more sensible approach of enlisting a lead horse that is unworried by the livestock to impart confidence to her horse.

■ Cause

Horses are far more sensitive to strange sights and smells than we are. Being prey animals their natural defence strategy is usually to run away, but on occasions they will defend themselves and in some cases attack something they view as a predator. In the case of dogs, many horses have had the experience of one running through or snapping at their hind legs, and so may naturally be wary of them.

■ Remedy

We are limited in how we can change a horse's intrinsic perception of other animals. It is sometimes possible to 'borrow' a donkey, cow, dog etc. to introduce to your horse or ask a farmer if you can graze him and his companion in a sectioned-off corner of a field containing cattle, for

Right: By introducing your horse to a dog in a controlled manner, you can aid the socialization process.

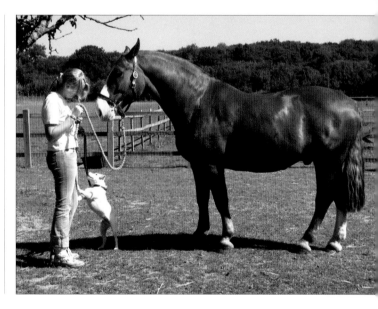

example. These will help to desensitize such an extreme reaction to the disliked species. However, each and every animal has its own smell and acclimatizing a horse to one cow does not guarantee that encountering cows in another field will not produce an acute response.

Case History – A Cautionary Tale

My lovely cob, Harry, was like a police horse – nothing would upset him. We could ride through fields of cattle and meet dogs while out with no problem at all. He was very used to my own dog and I thought it would be a good way to exercise both. The first time I took them out together was along a very safe track, away from all roads,

Above: Introducing horses to as many animals as possible in a safe, controlled environment will help to prevent a problem should you encounter the unexpected in an unfamiliar environment.

Above: A horse that has a bad experience with an aggressive dog at a young age may develop a fear of them that is difficult to eradicate even when the horse is mature and fully grown.

next to woodland. All was going well and my collie Jess trotted along behind Harry. I decided to canter and thought that Harry's attention would be focussed ahead of him but he was obviously concerned about the dog behind. He jumped about a bit and the next thing I heard was a yelp from Jess. I pulled up Harry and dismounted to discover that he had 'booted' her in the chest and she was bleeding and bruised. Jess and I were lucky on that occasion – it could have been fatal.

The dog recovered but I did notice that if any dogs ventured into Harry's field, he would sometimes charge at them full pelt with neck lowered and outstretched. I believe he saw the dogs as predators and consciously tried to keep them away from himself and his field companions.

■ SCARED OF WATER OR MUD

■ Problem

A seemingly irrational fear of a tiny puddle or area of mud can result in the horse backing off, leaping over, bucking or even climbing up or down a bank or ditch to avoid such a 'hazard'. The horses seem to forget they have a rider on board who can be dislodged or hurt in the process. They also often put themselves in further danger by leaving the pathway and jumping into areas which may expose you to more serious hazards like rubbish in a ditch or into the path of traffic. Added to this, a rider does not want to encounter difficulties during competitions, such as cross-country or Le Trec, where water obstacles or jumps are likely to be included.

Left: Water can hide unsafe footings, such as boggy ground or slippery algae-covered ledges, so understandably some horses view it as a hazardous environment to cross.

■ Remedy

Just as horses in the wild have had to get over the fear in order to cross rivers to new grazing, we can give guidance to our own problem horses.

• Allow the horse time to take in the obstacle and what we want him to do.

• Never bully the horse otherwise he will see the appearance of water as a cue for his rider to kick and antagonize him.

• Walk an experienced horse backwards and forwards through the obstacle in front of the nervous horse.

■ Cause

What seems to us to be irrational is, in fact, a reliable survival method. Horses will approach water to drink but not necessarily place a foot in it, unless confident they will not stumble in a crater or be sucked into a bog to drown. Moreover, if they do not even need a drink, why should they take unnecessary risks in this regard?

• Get the experienced horse to 'pick up' the problem horse and only then encourage him through.

• Hold the saddle or a chunk of mane in case your horse takes a leap, to avoid jarring him in the mouth which would further upset him.

Left: A horse will always feel more confident in the presence of other horses and this may be all it takes to make him take that first step.
Right: Leading a nervous horse mimics how horses would stay together in the wild when crossing unfamiliar terrain.

Above: Horses enjoy a splash around on a hot day – just be careful that your horse does not decide to have a roll!

Right: Getting a horse accustomed to the feel of a hose or being sponged should be accomplished in gradual stages.

- Take every opportunity to lead your horse on foot or follow another horse over similarly scary obstacles including puddles in your stable yard (which previously you would have most likely steered him around).
- Make or buy water jumps to practise with at home.

Many horses do not like the feel of water splashing on their legs (which is one of the reasons they tend to urinate on absorbent ground). Sponging and hosing the legs occasionally when grooming will accustom the horse to the sensation of splashing so that wading through puddles will seem less of a shock.

■ BOLTING TO OR FROM A FIELD

■ Problem

Not everyone is lucky enough to be able just to open a gate between their stable yard and the grazing beyond; it is often necessary to lead horses some distance to specific paddocks away from the stables. Attempting to lead a lively horse from the ground can be daunting and overly keen horses can easily drag or break loose from their handlers and bolt out of control. Horses can also misbehave when you are attempting to release them into the field, spinning round or dragging the handler before the halter is fully taken off.

Above: It is all too easy to be dragged or trampled as you attempt to walk your horse to his paddock.
Left: Confinement causes pent-up energy which may be released when the stable door is open.
Right: Try not to leave one horse in the field when moving several horses. Take the final two together to prevent distress and episodes of fence walking.

■ Causes

- The animal is upset by the movement of other horses – they want to stay with the herd.
- They are running to where they know there is food waiting.
- Pent-up energy or general exuberance.
- Lack of discipline.

■ Remedy

It is normal for horses to be affected by the movement of others around them as they would naturally choose to move together as a group – solitude means vulnerability in a wild environment. Some horses are pretty laid back if others are taken away (perhaps out for a ride), whereas others gallop frantically round the field calling desperately. A horse will only become calm if he is habitually exposed to this experience and learns that the other horses will eventually come back or that he will be taken to join the other horses. However, don't attempt this training if, in the process of learning, there is a risk that a horse or handler may be injured. It is far safer to ensure that there are always at least two horses remaining in the yard or field and that they are moved together. Many livery yards or riding schools make this a firm rule.

It is a common practice for feed to be left in each horse's stable ready for when it comes in from the field. Surely we can understand why a horse may want to charge back to the stable for this reward. If horses like this are fed in their paddock, then this urge to rush back can be lessened. Many establishments use grids of individual paddocks so that each horse (or those owned by a particular person) can be kept separately. In this situation, it is possible to feed safely in this way.

Above: Horses naturally choose to move together as a herd whereas we take and replace individuals as we wish which can be unsettling for them.

Safety

Never feed just one horse amongst a group of horses in a paddock or a fight is likely to break out.

Control is of utmost importance. Once a horse learns that he can get away from you, he will show you little regard. It is terrifying to be dragged by a horse or trampled over and very easy for a horse to overpower you. Firstly, **always** wear gloves with a difficult horse – these will give you more grip and prevent rope burns (I find leather gloves are best as they provide good grip even when wet as well as not restricting dexterity to undo buckles etc.). If your horse is prone to take off with you in tow, lead him in a bridle. It could be one kept specifically for this purpose, without a noseband and with a lead chain attached to the bit instead of reins. If you do use a halter, attach the rope to the side nearest you to improve leverage or, alternatively, wrap the rope around his nose for added security. Carry a whip – this is useful to put between you to prevent the horse from barging into you and can be pushed into his chest if he tries to overtake.

continued ➡

Left: Wrapping the rope around the nose or clipping it on the side rather than back of the halter will improve leverage and control.

BOLTING TO OR FROM A FIELD

Horses are clever and know how to take advantage. A horse that is terrible when with you may walk like a lamb with someone else due to their dominant and assertive aura or calm handling method. It often helps to get someone else to instil good manners in your horse – a large man sensitive to horsemanship is the obvious choice! Once the horse knows what is expected of him, then he should behave better with you. His improved behaviour should also increase your own confidence in your ability to handle him. If he pulls, make him stand, back him up or turn him in a tight circle. Insist he does what **you** want him to do.

If he has a tendency to rush to his stable, don't leave the door open for him to barge through recklessly. Instead, tie him up outside (use his headcollar not a bridle to tie up from) and pick out his feet, or change his rug outside the stable. Feed him

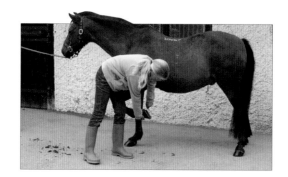

Left: If your horse barges into the stable anticipating that his feed awaits, discipline him to stand outside and wait calmly while you pick out his feet or rug him before allowing him inside.

a titbit if he stands calmly. You make the decision when he can go inside.

If he rushes to get loose in the field, make sure the gate is open wide enough for you both to pass through safely (without getting knocked or clothing/rugs getting caught up on the gate). Walk through and turn your horse back round to face the gate as you close it. Undo the buckle on the side of the halter but hold it with your right hand as you feed him a titbit with your left hand (this is so it appears that he is still restrained but the halter would actually fall off if he pulled back and ran away).

Above: Leading your horse in a bridle will give you extra control. Say 'Stand' and insist that he stands calmly while you undo him and then pat him or give the command 'OK' when you set him free.

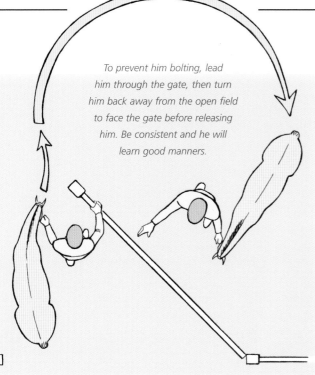

To prevent him bolting, lead him through the gate, then turn him back away from the open field to face the gate before releasing him. Be consistent and he will learn good manners.

Finally, remove the halter and give him another treat as a reward for staying. A rub on the head is the signal that he is allowed his freedom.

I do not recommend the use of ropes with 'quick release' clips which snap open if a horse pulls hard. Once a horse knows he can get loose, it is likely to become a habit and these ropes are only really suitable for tying up a horse in the safety of a stable or confined yard, not when you need to establish control over him.

As with most horsemanship issues, **fairness**, **calmness** and **consistency** will make a good partnership between handler and horse. If you watch a herd of horses they have their rules of who goes first and are quick to reprimand a member who oversteps the mark. This makes for a generally harmonious and safe environment.

Below: Once the gate is secured behind you, it is safe to remove the halter and allow your horse the freedom of the field.

■ BAD HABITS – PAWING THE GROUND

■ Problem

As soon as you arrive at the yard, your horse starts pawing at the ground to be let in or let out, or in anticipation of being fed. This behaviour usually gets more frantic as the horse watches your routine of preparing his feed. It is annoying because it makes you feel that you have to rush to get it done in order for the behaviour to cease. It also churns up the turf in the field and can unevenly wear down a horse's shoe/hoof on the side he favours for pawing, especially in a stable with a concrete floor.

■ Cause

In essence, this is natural behaviour. In the wild state horses would have more difficulty in sourcing their food and even water than they experience in a stable-kept environment. Although hardly dextrous, a hoof is quite a proficient tool for breaking ice on water or scraping snow or debris away from a food source. Quite often horses will scrape the surface of turf to

Left: This horse is pawing the ground as a display of irritation at being kept waiting. She is saying 'Let's get going'.

Above: A hoof is quite an effective tool at scraping a covering of snow and ice away to reveal the food or source of water beneath.

reveal the soil below to get at roots and as a source of minerals. They will also manipulate strange objects (and even prod the dung of other horses) in order to try and release odour to help identify them.

Survival in the wild depends on being resourceful at locating food and, like most other creatures, horses will sensibly display a degree of caution when exploring new things. Although a powerful animal in most situations, a horse is nevertheless vulnerable when his head is lowered to forage, as his eyes are at a level vulnerable to a nip on the face by an animal or snake hidden in the undergrowth.

In a domestic situation the horse that paws the ground is predominantly expressing frustration, generally in the expectation of food, but also sometimes when they are waiting, tied up and even under saddle when they are asked to stand but they wish to get going.

I have had many horses and ponies who have expressed themselves either before or during feed times. Some have pawed the ground while others held up a foreleg *(right)*, occasionally stamping it on the ground in excitement and perhaps as a threat to warn against any attempts to steal the food.

■ Remedy

It is not helpful to curb natural expressions as it is hard enough for a horse to try and communicate with us without suppressing this instinct. So it is better to try to have some understanding of what he is feeling rather than wanting a 'blank' horse whose spirit has been knocked out of him.

Usually, in regard to food, it is anticipation of a regular routine that has brought about this behaviour. The horse knows that, for example, when you get out of your car and unlock the feed room, soon he will hear the sound of the feed dropping into the bowls. You cannot blame him for showing enthusiasm!

You could try feeding him while he is still in the field, or leaving the food in his stable, ready for him as soon as he comes in, but he will soon learn the new routine so any cessation in the pawing behaviour is likely to be short-lived.

There are self-feeders on the market that work on a timer and allow the horse access to feed in his stable at a pre-set time. I am unaware whether a horse might paw the ground while watching the feed manger (when he thought it was due to open) but, in any case, the time clock can be altered each day to prevent this anticipation being satisfied. Thus there would be no obvious cue (like you arriving at the yard) to trigger an episode of pawing.

I would suggest you endeavour to live with this behaviour and view it as you would the antics of a young child who thumps his beaker on his high chair before he learns to speak and ask for more drink.

In order to mitigate hoof and shoe wear, it may be a good idea to lay rubber mats inside his stable door or where he is tied up. This will also dull the pawing noise which will benefit you and the people around you.

Above: It is a good idea to lay rubber mats over concrete areas to lessen concussion and wear if your horse insists on pawing.

■ WILL NOT LIFT UP LEGS

■ Problem

I just cannot get my horse to lift up his legs

Horses can easily plant their legs firmly on the ground and no amount of leaning on them or tugging on their feathers will work when you have 500kg (1100lb) of body weight against you. This is because horses naturally lean into pressure – unless they have been taught to yield away from pressure, you are going to lose the battle.

My horse is terrible with the farrier

As well as general fidgeting, horses have been known to rear up in order to get away from someone holding onto a foreleg and to kick out to try and shake off a restraint on a hindleg, so caution is advised with a nervous animal. It is not fair to expect your farrier to work with a horse that is not accustomed to having its legs lifted and it could potentially be a very dangerous situation.

■ Cause

The two problems tend to be connected. Sometimes a horse has not been taught to stand relaxed while his feet are picked up on a day-to-day basis and so finds doing this for the farrier (with all the additional unfamiliar sights and smells that this

Above right: Even at a very young age horses can be taught to overcome the natural feeling of vulnerability they experience when having a leg restrained.

Above left: A horse only feels safe with all four feet on the ground. It can be hard to persuade 500kg (1100lb) of horse to the contrary.

Right: Make life easier for your farrier by preparing the horse for his visits. Many farriers refuse to visit difficult horses unless they are sedated.

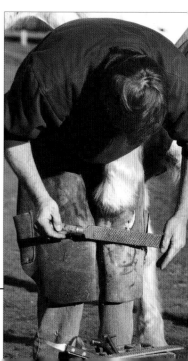

entails) doubly stressful. Alternatively, a horse may have had a frightening experience which he relates to being shod (e.g. lameness due to nailbind, rough treatment or simply being overcome by stress) and thereafter does not want to pick up his feet in any event.

Natural behaviour

Just as we would feel unstable standing with one leg restrained in the air, a horse feels particularly vulnerable in this position. Equines are prey animals who choose to flee from danger and trying to retrain them to accept situations placidly rather than escaping from them can be a difficult task. A horse needs to feel safe and secure to allow someone to manipulate a leg. It is helpful to have an assistant who can turn the horse's neck as if to lead him in a particular direction as this will shift weight off the opposite foreleg momentarily, so lightening it and making it easier to lift.

Remedy ➡

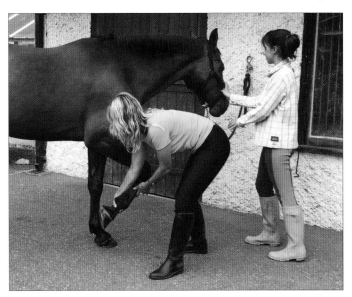

Above: Turning the horse's head to one side will shift his weight over to that side and so make the opposite foreleg lighter to lift.

Case History

My two-year-old Arab filly would bite me hard if I lifted up any of her legs. Being young, she was unbalanced and when any horse has a hoof off the ground they feel vulnerable as they are unable to flee. Additionally, she had suffered bouts of painful mud fever in the past and had become very sensitive to anything touching her lower legs. This cycle of fear of pain and fear of being put off-balance caused her to bite in an attempt to prevent her legs from being touched. In this case as the trust between us grew over time, she was more willing to be handled and rubbing her tummy *(below)* – which she loved – helped her to associate my touching her legs with a pleasurable experience.

■ WILL NOT LIFT UP LEGS

■ Remedy

Step 1 Make it a pleasurable experience. Make lifting a leg part of the grooming routine or bring him in from the field to check him all over for cuts or bites and perhaps give him a small feed at the same time. Do not prod or mither him, but stroke him smoothly all over his body and legs. Make sure he is standing squarely and in balance – you will not help his confidence if you try to lift one leg when he is already resting another.

Step 2 To pick up a foreleg, stand with your body level with his shoulder and facing his rear. Try leaning on his shoulder and just running your hands down the tendon from the back of the knee to the fetlock and gently lifting the pastern or his feathers. If that has no effect, pinching (firmly but gently) the tendon at the back of the leg should encourage him to lift it. For a hindleg, pinching the hock or the chestnut should have the same effect.

Step 3 Do not allow the horse to stamp the leg straight back down. Hold it still for a few seconds and gently put it down (do not just drop it). Even if he does stamp it down, try to keep

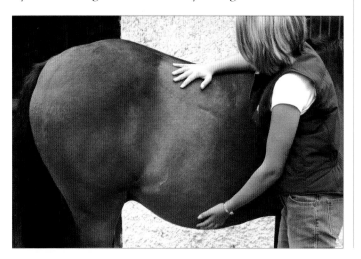

Left: Run your hands over the horse's body gently to accustom the horse to being touched all over.

Above: Support the hoof's weight before lowering it to the floor. Take care not to drop it abruptly.

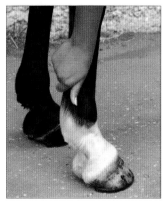
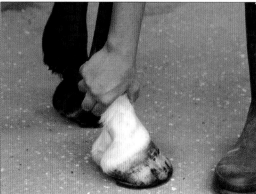

Above left and centre: Gently grip the tendons of the foreleg and lean into the horse to shift his weight over. Slide your hand down and lift from the pastern or the feathers.

Above right: Hind legs are often trickier and the horse may put up more resistance. Pinching the horse's hock can encourage him to lift that leg but take care he does not kick out suddenly.

hold of the leg firmly until you choose to let go. By letting go prematurely, you will reinforce the habit that he can evade you. Repeat this exercise and praise him as he improves. In time you should only have to run your hands gently down from the knee or elbow to achieve a lift.

Step 4 As he gains confidence, get capable friends (male and female) to do the same. Let them drive into your yard, slam the car doors and handle your horse. Let them pick out his hooves and tap the hoof wall and sole gently with a plastic hoof pick.

Above: Ask other people to handle your horse and pick out his feet. If he stamps it down, keep trying until they can dictate when the hoof is lowered.

Step 5 When the farrier comes, stroke and talk to your horse and perhaps feed him carrots or allow him to chew his leadrope if he is nervous. If he is going to be standing on the yard for some time, give him a haynet and perhaps switch the radio on softly if he enjoys it. An excellent way of increasing his confidence tenfold (and therefore stabilizing his behaviour) is to have another calm horse tied only a few feet away. The confidence this experienced horse exudes will reassure the nervous animal far more quickly than any anything we can do to help him.

If he will not stand still, do some exercises to focus his attention on you – back him up, push his hind quarters away from you, lead him, circle him and back him up again. Then ask him to stand and stroke his neck. Farriers are generally good horsemen and are very patient but I have seen them use their rasps and hammers to give a horse a wallop and this simply is unacceptable.

Ask your farrier to 'thank' your horse by giving him a gentle rub or a carrot afterwards. After all, it is in all your interests to make the experience as stress-free and safe as possible.

Artificial and Natural Aids
Humane twitch – Sometimes it may be necessary to use a humane twitch which releases endorphins to calm the horse.
Drugs – The practice of doping horses to enable them to be shod is a less desirable solution as it only masks the problem. However, it may be necessary in some very difficult cases and is preferable to neglecting the footcare of difficult animals.
Massage – Massaging tired legs with violet or lavender oil should help create trust between horse and handler and be a pleasant experience.

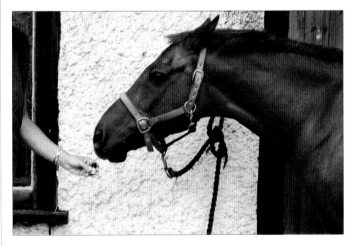

Camomile is a natural herb known for its calming effects and aromatherapy helps to de-stress a fearful horse *(above)*.

■ PLAY-TRAIN YOUR HORSE

Play-training helps with so many problems including horses that are bad to catch, bad to lead, or those that show lack of motivation, lack of coordination etc. and fixes your place as an Alpha figure in the herd hierarchy.

If you watch a happy group of horses in the field, they will delight the observer with fantastic dressage-type elevated movements, spins, flat-out galloping and sudden stops. An established group will have their hierarchy sorted out and the horses will be confident in how they interact with one another. How great it would be if you were regarded as an intrinsic part of the herd that your horse belongs to and for him to want to be with you.

Play is such an important part of natural behaviour. We can use and mimic this and enjoy a more light hearted approach to horse training that is fun and which helps to cement your relationship with your horse. Here's how…
You will need:

• An enclosed area (a small paddock or sand school is ideal).
• Your horse should be wearing a headcollar (ideally with 30cm/1ft of rope or baler twine hanging from the ring behind his chin).
• A rope at least 2.4m (8ft) long or a lunge line.
• An expressive voice!

Note: Play-train **after** the horse has had some free time to chill out. Do not try to do it with a stressed animal or one who has come straight from the confines of a stable or he will see it as work rather than play.

Set your horse loose into the enclosed area. It should not be so small that he cannot canter freely but equally not so large that you have to trudge miles backwards and forwards after him. Leave him alone for a minute or so to loosen up and sniff around or perhaps roll on the ground. Enter the area and approach and stroke the horse. Show him your rope and allow him to sniff it.

Above and right: Start by draping the rope over the horse and rubbing him with it. Get into a rhythm of throwing it over and pulling it slowly off. You do not want it to be seen as something to fear. A nervous horse can be desensitized also by draping the rope round each leg.

Above: 'Weeeeeee' let's have some fun! My horses buck, snort and prance around. You are not trying to wear them down, but become an Alpha horse.

Above left and right: An Alpha horse can move other horses away just by poking his nose or swishing his tail. Backing him up and moving his quarters away will raise your status; your commands will become minimal with practice.

Drape it over him and rub him with it. Walk away from the horse.

Here is where you start to look completely insane …… Start spinning the rope vertically and in an upbeat and excited tone begin to squeal 'Weeeeeeee, weeeeeeee' as you continue spinning it. It sounds crazy to us but it communicates pleasure and joyfulness to the horse and perhaps is similar to a horse squealing in delight. Your horse will probably career off and look at you in amazement! Move around the area, facing the horse and continue the spinning and squealing. Allow him only a few seconds to stop and look at you and then encourage him to move off again. Throw the end of the rope towards him if he gets stuck in a corner to make him move off. The difference between this and 'Join Up' is that you do not want to exhaust him or make him submissive, so you do not persist with making him move off until he has had enough, but just to get him excited and agile.

Hold the rope limp and approach and stroke him. Jump up and dart towards his quarters – watch him move away! Then run back and perhaps sit on the floor and wait for him to approach you (keep your eyes down). Stand up slowly and rub him where you know he likes it. Hold the short rope from his headcollar and make him go '**Back**' away from you and halt, then (still holding his head) move his quarters away from you and then stroke him. Repeat the '**Back**' movement then say '**Over**' as you move him away on the other side and then stroke again.

Back away from him and start spinning the rope again and squealing so he runs off. He may buck and snort! Approach him again and stand alongside him and, lightly holding the short rope from his headcollar, encourage him to trot next to you. Let go if he goes faster than you. Try to be just ahead of him and trot together for several strides and then throw back your arms shouting '**Whoa**'. He should bounce to a stop.

It may take a while for your horse to understand this new game but once he does he will be following you about like a lamb. The games can be enhanced to include objects such as poles, corridors, jumps etc. An inexperienced horse will be greatly inspired with confidence, rather than quaking with dread at encountering new objects, if you are taking the same steps next to him rather than sitting on top, kicking and shaking the reins at him! These games can be continued throughout your partnership with you horse. Spending time together creates a special bond, so when you are short of time to ride or just want to have fun, play-train your horse instead.

CHAPTER TWO

STABLE VICES AND MANAGEMENT PROBLEMS

The dictionary definition of 'vice' is '*an evil or immoral practice or habit, wicked or evil conduct or corruption or a flaw or failing or physical defect*'. It is sad that so many horses are labelled as having a 'vice' whereas more often than not the problem behaviour has been caused by the mismanagement and inadequate skills of their owners! Unfortunately, any bad habits learned by horses are quite tricky to 'un-learn' but some of the common problems are addressed in the following pages and practical solutions suggested.

■ BITING AND CHEWING

■ Problem

Biting is a common problem which causes scores of 'non-horsey' people to think the rest of us are mad. We have probably all been warned, 'Don't walk behind a horse or it may kick you and don't go near its head or it will bite you!' at one

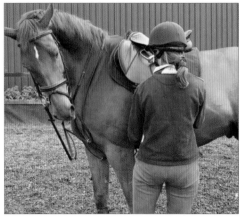
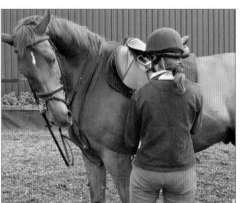

Above (left to right): Biting when the girth is tightened is a common problem. Adjust it in stages and avoid pinching the skin. Shorten the outside rein to keep his head at a safe distance from your vulnerable spots!

time or another, probably when we were children. Sadly it is probably the safest option for those without horse sense (or quick reactions). It can be made all the more alarming when your seemingly angelic animal does a Jekyll and Hyde act and takes a chunk out of an unsuspecting friend holding him for

you as you get the tack, or perhaps someone walking past his stable. It then falls to you, as the owner, to make the apologies or pay for the ripped jacket to be repaired – your horse certainly won't.

Left: Horses that are depressed by their confinement in a stable may lunge out at passers-by.

Biting over the Stable Door
Attention Seeking
Some horses crave attention and they find even the negative attention of a person shouting at them preferable to being

ignored for hours. This frustration manifests itself in a horse that lunges at passers by over the stable door.

Negative Association
Far less common is a horse that actually hates the sight of people. This can come about because they associate people entering their 'safety zone' (stable or personal space) as being a precursor to work or some other unpleasant experience.

■ Cause

From a horse's point of view biting and nipping are part of their repertoire of language. Gentle nipping (using the lips or gently holding the skin with the teeth) is generally seen as a friendly or playful gesture between horses. They do this when grooming each other, when saying 'hello' or 'play with me'.

Colts, fillies and stallions will be the most prolific biters. They see us as part of their herd and this is natural behaviour for them, indulged in as play or as a statement of power in the case of stallions who should only be handled by the most experience horsepeople. Remember also that the teething of

Stable Vices and Management Problems

Above: Horses will naturally nibble when grooming each other and do not understand that we may not appreciate the same when we groom them.

youngstock goes on for many years and this can cause discomfort which may be partially alleviated by chewing.

Biting and lunging with the neck outstretched is normally a defensive action rather than sheer aggression. It is generally done out of anger, pain or fear and is usually forewarned by aggressive body language. Between two horses it serves to drive the other horse away from its personal space and may be used to keep them away from a source of food or a vulnerable foal.

Horses use this same language in a domestic situation. Some do not like too much fuss and may simply be saying 'Enough' to a long grooming session, for example, or showing possessive behaviour over a bucket of feed. Unfortunately many people are blind to more subtle indications like a swishing tail or flattened ears and biting is therefore the horse's means of SHOUTING his displeasure. This, unfortunately, leads to horses and ponies learning that aggressive behaviour makes humans back off and it is easier to bully than be bullied.

When we see horses fidgeting when being shod or rugged, it is helpful to try and imagine ourselves in their position. In the human world we can explain to each other what we are about to do, but it can be a challenge for a horse – who has to deal with being on the receiving end of items of tack, sprays and lotions being administered and limbs being pulled about – to

Some Specific Causes of Biting	
Bites when girth or rugs are being done up or being groomed.	Discomfort, pain, fear (not only of the objects but of their association e.g. being ridden, leaving companions etc.).
Bites people passing stable or when tied up.	Attention seeking, lack of mental stimulation, lack of physical activity, distressed at confinement in stable, lack of forage.
Bites while eating or bites when given treats.	Protective of food, hungry, disrespectful of people and their personal space.
Bites in a stressful situation e.g. when being shod, faced with a difficult task or object to overcome.	Seeks reassurance by touching and holding onto things with the mouth, lack of confidence.

keep calm when he has no power of control. Chewing the leadrope *(left)* or somebody's coat is often a displacement activity (a bit like a child sucking and nuzzling a security blanket) to reassure and comfort. In a natural environment a horse would simply move away if he found that something disturbed him, but in a confined environment he does not have this opportunity.

Remedy ➡

■ BITING AND CHEWING

Problems Caused by the Rider/Handler	
Lack of horse sense and understanding	Far and away the most common cause. If you cannot recognize subtle signs being displayed by your (or anyone else's) horse, the horse has no choice but to make them more obvious.
Not thinking about the cause	Think back to the occasions of biting. What is your horse trying to tell you? Is he getting enough turnout, enough mental and physical stimulation? Does his tack hurt him or are you simply being rough when grooming or perhaps pinching his skin or pulling his whiskers by accident when tacking up.
Asking too much from your horse/pony	A horse is a living animal and not a machine. They get tired and have 'off days' and will soon get stale and resent being ridden if their workload is not varied to maintain their enthusiasm.
Feeding titbits ad lib	If you carry mints or treats in your pockets you are encouraging your horse or pony either to nudge you incessantly for one or try to take a chunk out of your pocket himself!

■ Remedy

Ultimately it falls to us to redirect bad horse behaviour. Here are ways to help achieve this end.

- Be assertive at all times. It is usually nervous or inexperienced people who get bitten. Being assertive certainly does not mean being aggressive, just making your wishes clear and having the ability to keep a horse out of your personal space and focused on your commands.

- Be aware of your horse's moods and tell anyone handling them if you suspect he may be grouchy – we all have off days.
- Be sympathetic to your horse in nervous situations. Stroke him and talk to him and if he needs something to chew to reassure him, allow him have a chew on the leadrope. Touch is a fundamental part of life for nervous horses lacking in confidence.
- If he bites when tacked up, have his teeth examined by a

dentist *(above)* and the fit of his tack checked. Make a conscious effort to be gentler when you touch him, whether it is during on the ground grooming and tacking up, or with your leg, seat and hand aids when riding.
- Reassess his management routine – perhaps he needs more turnout, more exercise or playtime with his companions (human and equine).
- Do not carry treats on you (unless you know your horse does not have a problem with them). Only feed treats as a reward, for example after a ride as a 'thank you'. Dissuade other people from feeding your pony treats over the stable door.
- Check that your horse is getting enough forage to satisfy his hunger and also his desire to chew. Long periods without

Stable Vices and Management Problems

forage are not good for mental or physical health.
- If your horse goes to bite you, do not hit him round the head. This just turns the exchange into a battle or could make the animal headshy.
- Block any attempt at biting (if you are quick enough!) with a plastic curry comb or your whip. Whether or not he actually bites you or your clothing, immediately let out a piercing scream, wave your arms and scare the hell out of him instead! (Warning: he may rush back and break free from the leadrope so make sure the area is secure).

Above and below: A grooming brush or the whip you are carrying can be used as a good instrument of defence to block a sudden snap of teeth.

- If he lunges at you in the stable or field, send him away from you. Adopt a powerful head-on stance with shoulders and arms raised and eyes locked on him. For safety carry a schooling or lunge whip and tap it hard on the ground in front of you and march towards him. He must learn to keep an appropriate distance from you, unless YOU invite him in closer.

Dietary Remedy

If he is stabled, perhaps his diet could be altered to be almost entirely made up of high fibre forage without hard feed depending on energy intake. For example, the forage could be varied in different feeds of meadow hay, thrashed grass, barley straw, alfalfa and various short chops with the addition of succulents such a carrots and non-molassed sugar beet, along with vitamin supplements if required. This will be far more satisfying in terms of chewing and creating a full feeling 'than if the same calories are obtained from adding hard feed in the form of cereals. It may well stop any problem with snapping over treats and the horse being aggressive at feed time.

Alternative Remedies

For horses that are miserable being stabled try fresh or dried rosemary and camomile added to the feed. Aloe vera juice, rose hips and thistles are good tonics and skin conditioners for sensitive horses, and again can be added to the feed.

Above: Plenty of forage, treats and succulents scattered in the stable will satisfy the natural grazing and foraging desire.

Above: Herbs can be offered fresh or dried to promote good health and a feeling of well-being.

■ KICKING THE STABLE DOOR

■ Problem

As soon as you walk into the stable yard, you hear the incessant bang, bang, bang as your horse bashes the stable door with his knee or scrapes it with his hoof. This results in you having to shout at the horse and this situation can soon turn into an irritating cycle of him banging and you shouting which tends to annoy other people on the yard.

■ Causes

This behaviour is often exploited by more intelligent horses who, quite understandably, have found a way of expressing themselves. Often you will see that this behaviour occurs when they can hear their feeds being made up and they are basically saying 'hurry up', or when they are eager to be let out into the fields. Often the yard is totally peaceful and all it takes is for them to hear one footstep before the riot commences.

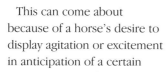

This can come about because of a horse's desire to display agitation or excitement in anticipation of a certain event (e.g. feeding, being turned out). Pawing may not be enough to transmit a satisfactory visual or audible message to his owner if the horse is in a stable with the door shut or has a layer of bedding to mask any sound. Kicking or kneeing the door is in some ways a similar display to a young child thumping the table or fidgeting in his highchair when he cannot reach a fallen toy. Calm is restored when he finally gets his reward.

This is one of the drawbacks of keeping horses

Above: Banging on the door is only his way of communicating with you – 'Hurry up with my food'. Horses do not understand etiquette!

stabled and leading a life that usually involves a regimented timetable of feed, hay, exercise, hay, feed etc. This is fine for their physical well-being but unfortunately provides very little else to occupy the horse mentally. This means the horses focus on their next stimulus and hang on every sound and movement in the yard. By comparison, field-kept horses frequently display a total nonchalance when you arrive at the gate!

■ Remedy

- Firstly to address the actual noise issue. A simple countermeasure is to attach a large piece of thick carpet to the back of the stable door (screw the carpet in place or use fencing staples as nail heads would too easily be pulled through the material). This also serves to limit any bruising to the horse's legs.
- Secondly, address the cause because if you merely curb this type of behaviour it will only result in the horse seeking another outlet of display and fuel his frustration still further. Allow the horses to interact more – ideally in the field where they is room for both closeness (e.g. mutual grooming) but also space for solitude if required (rather than a potentially stressful environment of being stabled next to an animal they may dislike). Field-kept horses are much less reliant on routine as there is so much more to occupy them.

Right: Stabled horses are completely reliant on their owners and focus on any noise and movement in the yard, hoping for attention or food.

Stable Vices and Management Problems

- For stabled horses, fix a grilled partition on one or more sides of the stable so they can interact with their neighbours (ideally their pair-bond). Stable cats and even chickens help to keep the yard active when people are not there but they are relatively quiet and less stressful to be around than a barking dog, for example.
- Ad-lib fibre – if horses are not able to graze pasture (which will occupy them both mentally and physically), give them a variety of forage placed around the stable in small holed haynets containing perhaps hay mixed with barley straw, and a bucket of alfalfa or a grazing block. It is unnatural for horses to go for long periods of time with no food and most horses can perform just as well 'off grass' or with ad-lib good quality forage (and will be happier mentally) than those who receive just two meals a day of cereals and a haynet.
- Give the horse more to do. Often these horses have a high intelligence and enjoy having a job to do, so increasing their workload should help them focus their mental energy and enjoy the time they are stabled for rest and relaxation, rather than seeing it negatively as confinement.

Tip

A chain, breast bar or webbing barrier will make the horse feel less confined and remove the solid barrier which he has previously kicked.

Above: Animals such as cats, chickens, goats and alpacas can be a valuable contribution to the stable yard and provide company and activity while being more peaceful than noisy people! Dogs can form close attachments with horses but a dog barking incessantly will be stressful for the horses.

Right: Using a chain or breast bar will prevent the horse from banging the door. The rubber matting in this stable also inhibits the horse from pawing the ground noisily for attention. With a chain, more light comes in making the stable seem less claustrophobic.

◼ EATING BEDDING

◼ Problem

As with crib-biting, weaving and wind sucking it can be worrying to see a horse displaying obvious displacement behaviour which has been learned as a 'release' and established itself a habit. Eating bedding should only be a worry to us in terms of the suitability of what the horse is actually consuming.

◼ Causes

The action of foraging and chewing is wholly natural and from an owner's point of view eating bedding is the lesser of many evils in terms of the horse seeking a release from the constraints of confinement. A horse given ad-lib forage will be very unlikely to display this behaviour and so it may be that the ratio of cereals to forage is wrong – by eating bedding the horse is trying to self-medicate.

This trait is not only displayed by horses who are stabled for long periods. Horses on lush paddocks may come in to the stable and start to eat their bedding. Forage is by far the most important component in a healthy equine's diet and is necessary for the gut to function correctly. Eating bedding or wood should ring alarm bells that a greater fibre content in the diet is required. This does not have to mean that your horse will put on weight.

◼ Remedy

Fibre in the form of chopped straw (not the treated type used in bedding), long stem straw, thrashed

Left: A horse gorging on bedding risks impaction colic and this shows that his mental and physical need for adequate forage is not being addressed.

(de-seeded) hay and even tree branches (check these are not poisonous) will help satisfy the horse's chewing reflex and promote mental well-being, without undesirable weight gain. Horses on perfect lush paddocks should be offered roughage as the water/fibre content of grass changes through the seasons and only a few horses enjoy the luxury of having a gourmet selection of tree species to browse to raise their roughage intake.

Below: Suitable hedgerow branches should be offered to all horses without varied grazing. Dispense forage and cereals throughout the stabled period rather than in 'set' meals to keep the gut full and the chewing instinct satisfied.

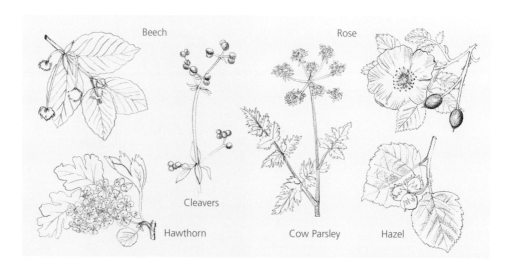

Beech
Rose
Cleavers
Hawthorn
Cow Parsley
Hazel

■ RUG BITING AND DESTRUCTIVE HABITS

■ Problem

It can be very annoying after you have spend your hard earned cash on a new rug to come back to the yard and find the rug torn to shreds, or a new hay rack you have spent several hours hanging up now swinging on one screw. Rug biting and destructive behaviour such as ripping wood trim or haynets off stable walls may not be solely displacement behaviour to combat boredom or lack of fibre as outlined in the subsequent sections about stable vices.

■ Causes

If it is not a boredom/frustration issue, then such behaviour may be more directly linked to the actual items involved. Horses are able to regulate their own body temperatures brilliantly by the raising or flattening of the hairs on their skin and by growing or shedding winter and summer coats. We rug horses with the intention of keeping them warm, dry and happy, and this usually is the case. However, by doing this we take away their innate ability to modify temperature and make them reliant on us to do it for them. Even the best owner in the world will have come down to the yard to find their horse sweating or shivering on an occasion when we have been unable to act quickly enough to a change in temperature.

Rug biting and tearing could simply be your horse's way of trying to get the rug off as he becomes uncomfortably warm or trying to get to an itch which is swathed in several layers of wadding and material and therefore increasingly aggravating.

In an attempt to have a good scratch of his backside, your horse may dislodge a protruding bracket or hay rack – half a tonne of incessant itching can shift even the biggest screw or best carpentry work!

■ Remedy

- Always look at the overall picture and routine in case the behaviour is caused by general discontent in his environment.
- Only use rugs when necessary – not simply because the owners around you may do so. Pay regard to your horse's breeding or type and what (if any) clip he may have. Horses generally cope well even in very low temperatures if they

Left: Many stabled equines are over-rugged and underfed to keep a desired weight. They would be happier and healthier if fed larger quantities of hay and allowed to let the digestion of this keep them warm and mentally satisfied.

have plenty of forage as heating fuel but can rapidly become miserable and lose weight if there is a prolonged wet spell which saturates their hair so it cannot maintain heat.

- A natural or man-made shelter is more important than a rug both for staying dry and providing shade in hot temperatures. The overriding benefit is that a horse can choose when to walk in and out of a shelter when needed – but not so a rug!
- Any rugs used should fit the horse well.

Below: The horse is highly efficient at regulating temperature by erecting or flattening hair. A rugged horse has no such control over his temperature.

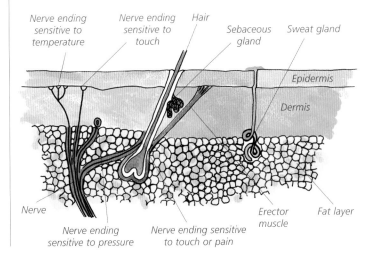

■ STABLE VICES

■ Problem

Watching a horse displaying stable vices such as weaving, box walking, crib-biting and wind-sucking is distressing as the actions are repetitive and the behaviour is compulsive. Even if you try to interrupt the bad habit, often you only cause a momentary cessation and then the horse resumes what he has been doing as soon as he is able to. It is often a worry that other horses on the same yard will copy the behaviour and, as horses evidently derive some comfort from it, once learned these vices are extremely difficult to eliminate.

Weaving is generally seen at the stable door. The horse will shift his weight from side to side and weave with his head and neck as a therapeutic release from a stressful or emotive situation (e.g. watching other horses leaving the yard while he is confined or hearing his feed being made up).

Box walking is similar attempt to release frustration (like a caged tiger at a zoo), where the horse will pace round and round the

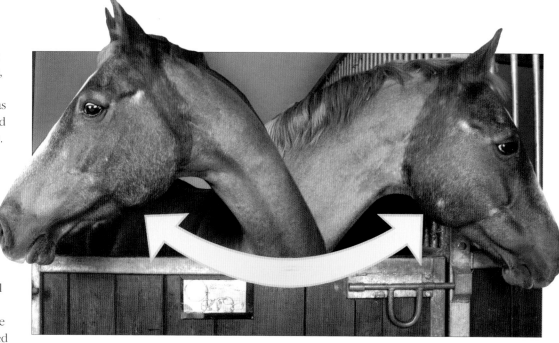

Above: A horse weaving is a distressing sight – like a caged tiger. Don't just treat the symptoms – it is your responsibility to improve conditions to alleviate the cause.

Below and right: Wood chewing makes for an unsightly stable yard but we really should look at our management systems to ensure they are not the cause of the problem.

Left: Anti-weave grilles are common but simply make the stable even more prison-like. Some horses really cannot tolerate being stabled for long.

confines of his stable and can become very anxious and sweated up. Often horses that have been prevented from weaving (by the installation of anti-weave grilles in the stable door) will actually modify their behaviour to include box walking.

Crib-biting is where a horse grabs hold of a solid object with his teeth, such as the stable door or a fencing rail, and will arch his neck and pull back on his incisors. Some horses will chew wood and others will gulp in air, known as wind-sucking. Apart from unsightly damage to any woodwork, it can contribute to uneven teeth wear and also susceptibility to colic. Initial findings from studies of colic in horses seem to point at wind-sucking horses being up to 50 per cent more likely to

suffer the more severe forms of colic requiring surgery, such as entrapment and twisted gut which can be life-threatening.

Although horses perform this action throughout the day and night there are often regular triggers that start them off, such as finishing their bowl of food or horses moving on the yard.

Causes

- The horse has been prevented from performing a natural behaviour (e.g. adequate grazing or social contact) in their past or present environment. The stable vice provides some appeasement of the frustration of not being able to carry out the desired natural behaviour. Even if his lifestyle changes and the cause is eradicated, he is quite likely to continue displaying this behaviour – a bit like a child sucking its thumb for security.
- Restricted foraging ability or too little fibre in the diet.
- Endorphins, pain-relieving substances similar to morphine, are released during the behaviour so the horse can become addicted to the feeling it produces.
- Stress from separation from other horses – lack of social contact.
- Displacement therapy – there is nothing else to do!
- It can be an inherited condition as foals have been observed crib-biting without any apparent influence from other horses or poor management.

Above: Many vices are displacement behaviour when a horse is prevented from performing natural behaviours while stabled. Chewing and foraging plus social contact are the mainstay of a happy horse's life – not confinement in a stable.

Remedy

■ STABLE VICES

■ Remedy

Although there are contraptions and foul-tasting preparations on the market to prevent a horse from crib-biting and grilles to thwart weaving, I would never advocate the use of such items until you are satisfied that you have done everything else possible to address the possible cause (e.g. stress or confinement). Scientific studies have proven that horses that are physically prevented from carrying out these behaviours (which are already displacement activities because they are being deprived of other needs) suffer further anxiety and will either establish alternative undesirable activities or become generally depressed or bad tempered. If the restriction is removed, they will frantically strive to resume the old behaviour– a bit like a smoker lighting up after he has been prevented from smoking.

Treat the cause, not just the symptoms, then you will know that you are providing an environment which satisfies the horse's most inherent needs.

• Feed as much of the horse's diet as forage – ad lib if possible. This will prevent acid build-up in an empty stomach and should dramatically reduce stress and help those that crib-bite and wind-suck.
• A bed of barley straw will allow the horse to continue to 'graze' even when his haynet is empty.
• Tree branches, carrots and some loose pony nuts scattered around the stable will allow the horse to indulge in natural foraging behaviour not normally offered when he is being fed his usual meals.
• Provide company – grilled partitions between stables are good for enabling horses to touch and smell each other and reduce stress. Make sure, however, that they will not be intimidated when eating.
• Loose barn stabling gives horses freedom but keeps them indoors.
• Increase turnout.

Above: People prefer bedding that is inedible but in many cases straw bedding could be used to supplement the diet and provide natural 'grazing'.

Above: My horses can choose whether to be 'in' or 'out'. I never feel guilty that I am not there early in the morning to let them out – do you?

Stable Vices and Management Problems

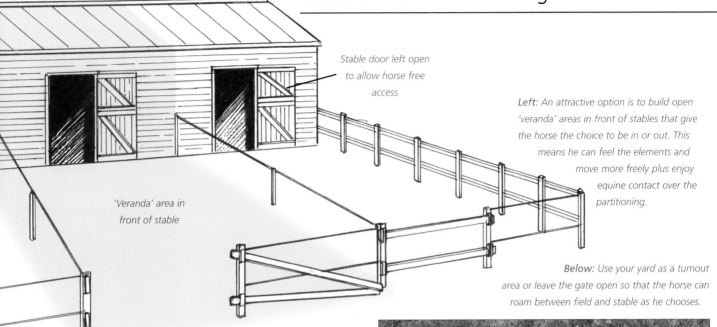

Stable door left open to allow horse free access

'Veranda' area in front of stable

Left: An attractive option is to build open 'veranda' areas in front of stables that give the horse the choice to be in or out. This means he can feel the elements and move more freely plus enjoy equine contact over the partitioning.

Below: Use your yard as a turnout area or leave the gate open so that the horse can roam between field and stable as he chooses.

It is relatively easy to build open 'veranda' areas in front of traditional stables. This serves the purpose of doubling the size of the stable and allows horses to get fresh air and touch one another across the partitions.

Above: A superb facility for rotational release. The horses take turns for a 'night out' and provide contact and stimulation for their stabled companions.

Another alternative which is particularly useful on a livery yard is a rotational release system. In this approach, the majority of horses are stabled as usual, the barn door or the yard gates are firmly shut and one horse is left with his stable door open. This allows him to roam around the yard or barn. Not only will he enjoy freedom but he will also provide social contact and a distraction for those still in their stalls. In an average yard your animal may only have one night out every week or fortnight but on the remaining nights he will have equine contact and be entertained by the other horses released.

continued ➡

■ STABLE VICES

**SEVEN STEPS TO KEEP A STABLED HORSE
MENTALLY HAPPY AND HEALTHY**

1 Provide a view. Movement and activity in your horse's field of
vision will help to stimulate his mind. It can be human, animal
or even farm machinery
working in the next field.

*Left: Horses are highly intelligent
and need mental stimulation,
forage and company.*

2 Provide equine contact. If
your horse is kept out in a
field, make sure that he has
company that he can touch. If
this is not possible, turn him out somewhere that has horses
in neighbouring fields so they can communicate and smell
each other. Stabled horses should be able to see other
horses. However, this is not as rewarding as being able to
groom and socialize freely, so make sure that they are turned
out frequently with at least one other animal, and allow them
to build friendships.

3 Provide long-lasting forage. Horses' digestive systems were
designed to take in small amounts of food over 24-hour
periods by constant grazing, rather than having one or two
large meals a day as we do. A small-holed haynet and several

small feeds a day will be
easier for the horse's gut to
digest and will keep him
occupied for longer while he
enjoys his food.

4 Provide a task. Horses are
intelligent and have to search

*Left: Stable toys and slow-release
feeders can go some way to occupy
the horse and satisfy foraging
behaviour in a stabled horse.*

for food in the wild. To stimulate stabled horses, leave treats
hidden under clean straw to be found later or hang a swede
from the rafters which will take time to eat. There are toys on
the market which incorporate licks and others that the horse
must roll on the ground to release pony nuts or treats
hidden inside. It all helps to keep him mentally stimulated.

5 Provide entertainment. You
can use a horse toy *(right)*
to distract a solitary
animal, but other animals
can help to provide
entertainment in the form
of something to watch and
even play with. A stable cat
can act as companion and
rat-catcher combined!

6 Provide atmosphere. A
lone horse in a quiet atmosphere will retreat into itself so
provide an atmosphere if there is nothing else to stimulate
the horse. Horses enjoy music, so leave a radio on for an
hour or two on a 'suitable' station. If you have no power at
the stables, there are now radios on the market which need
no electricity or batteries – they run off solar power or by
being wound up.

7 Provide comfort. Apart
from food and water, a
stabled horse needs
suitable bedding to help
support his weight on his
legs or to allow him to lie
down or roll safely *(right)*.
But remember that horses
are not designed to stay in
one place for hours on
end, so unless injury
dictates otherwise, they must be allowed to exercise freely to
prevent fluid retention and swelling in the legs, and general
stiffness.

I have done everything – my horse has forage, freedom and friends but still crib-bites – what can I do?

If the behaviour continues, it is likely to be a learned response (which can even begin as a foal) or otherwise a habit that stems from **previous** restrictive management which has not been eliminated (similar to a young child sucking his thumb for comfort and this continuing beyond infancy). In the case of crib-biting or wind-sucking it is advisable to attempt to break the habit for reasons of its possible link to colic, which can be life-threatening. The first step should be to make any fences and surfaces less appealing to touch by painting them with foul tasting anti-crib gel or paste. Plastic pointed strips usually employed to deter cats (and burglars!) from fences can be used on the stable door and window sill.

Crib-biting collars should be used with care and removed frequently to check for rubbing. Make sure they have leather straps which will break if the collar catches on something. There are some more humane examples made of a leather with a sheepskin lining which are equally as effective in preventing the horse from arching his neck to crib-bite but which certainly seem far less severe in their action than the old traditional collars.

Right and below: If you have a clear conscience that you have addressed management issues but the crib-biting continues, then a humane leather collar will help break the vice cycle.

Below: Look beyond a smart rug or a spotless stable to assess if you are a good horse owner. Think of life from the horse's point of view rather than your own!

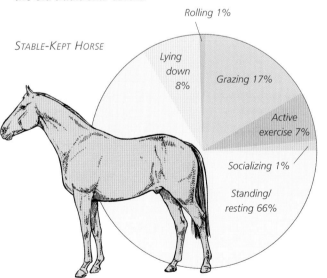

STABLE-KEPT HORSE

Rolling 1%
Lying down 8%
Grazing 17%
Active exercise 7%
Socializing 1%
Standing/resting 66%

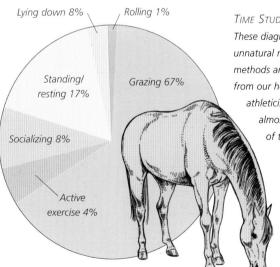

Lying down 8%
Rolling 1%
Standing/resting 17%
Grazing 67%
Socializing 8%
Active exercise 4%

WILD/NATURALLY KEPT HORSE

TIME STUDY
These diagrams illustrate just how unnatural modern horse-keeping methods are. We expect so much from our horses in terms of athleticism and yet they can be almost immobile for two-thirds of the day.

■ DIFFICULT TO RUG

■ Problem

It is often necessary to use a rug on a horse, for protection from the weather, for warmth if he is clipped or to keep him clean while travelling. It can be irritating if you have a fidgeting horse who circles his stable to avoid you or pulls back when tied up every time you approach with a rug. Accidents can happen if you are part way through doing up a rug and the horse gets away – the horse may get caught up in trailing straps or the rug slip sideways causing the horse to trip and panic further.

■ Cause

Some horses are naturally more spooky that others and will panic if you change even the colour of their stable rug! Rugs are difficult to manoeuvre smoothly onto the horse so it is quite common for a strap to slap them as you throw it over their backs. Ill-fitting rugs restrict movement and can rub areas of the coat bald creating sore patches. A previous incident of getting a leg caught up in a rug strap can also make horses wary. Existing saddle sores or insect bites will be aggravated further from the heat and friction of wearing a rug. A horse wearing a rug cannot regulate his temperature himself, as he would with his own long coat, and so may become uncomfortable when wearing one. Sweating under a rug is uncomfortable and upsetting for a horse but unfortunately few owners can be on hand to check their horse's skin temperature every hour or so!

■ Remedy

If you have a particularly spooky horse begin by approaching your horse with a saddle cloth and elastic surcingle. Show him the saddle cloth and let him sniff it. Rub it slowly along his neck and place it gently on his back. Hold it in place with the surcingle so he can get used to it. Progress to using his rug but firstly **learn how to put on a rug correctly** rather than just throwing the whole bundle on his back in a heap. Folding the rug means that a neat square (not much larger than the saddle cloth) is placed on his back and this can be gently folded out in stages. The straps will be under control rather than dangling all over the place and making him jump.

If he starts to panic, stop and wait until he is calm and accepts it on his back. You can use the surcingle to keep it in place during each stage of unfolding the rug, to prevent it falling off if the horse moves around. Check the fit of the rug, not only the overall length and depth but the tightness of the breast and belly straps. It is a common mistake that the top breast strap is tightened too tightly which makes it uncomfortable for the horse when he lowers his head to drink or graze. Also check that any leg straps do not hang down too low. When removing the rug, re-fold it inwards in a similar fashion starting from the tail end (unless your rug has a tight fillet string where it will be necessary to slide the rug over his quarters to remove it).

Left: You can get a good idea if your horse is warm by feeling the temperature around the base of the ear. This part of the horse's body is easily accessible even if he is wearing a rug.

Right: The area of skin behind the elbow should feel warm, but not sweaty, to the touch for the horse to be feeling comfortable.

How To Put On A Rug

1 Fold the back and front halves of the rug inwards until you are left with a panel not much wider than a saddlecloth. Ensure all the straps are tucked inside.

2 Unfurl the front section ensuring the hair remains flat underneath.

3 Fasten the chest buckles first and allow room for when his head is down grazing.

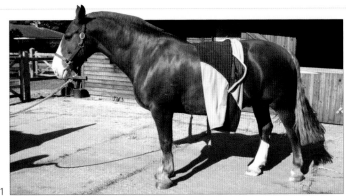

1

2

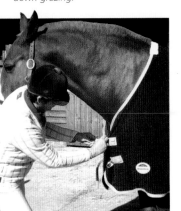

3

4

5

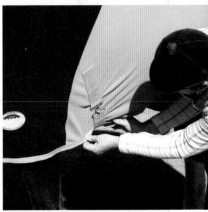

6

4 Work backwards and open out the rug carefully.

5 Fasten leg straps. They should loop through each other to avoid chafing and be high enough not to get caught when the horse lies down.

6 Fasten the belly straps in a diagonal – left strap across to right buckle, right strap into left buckle. Allow a slack of about a palm's width under the belly.

7 A well-fitting rug will not rub or slip backwards.

7

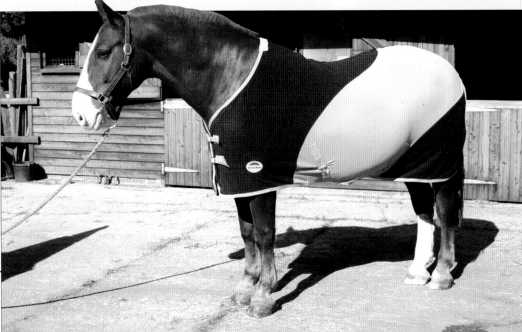

■ DIFFICULT TO BANDAGE

■ Problem

I am not a fan of boots and bandages being used by the average person on general pleasure horses as I believe it makes them too reliant on the artificial support and so inhibits the natural strengthening of the legs as they are put under more strain.

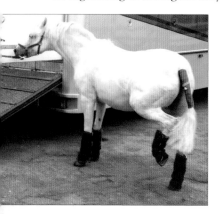

Above: Fasten leg bandages tightly to avoid a bandage slipping down as the horse walks or he kicks out, which could cause him to panic.

However, a horse's conformation or way of going may dictate that, for example, bell boots need to be worn to prevent over-reaching injuries or shoes being torn off, or tendon boots may be useful to assist on a particularly strenuous occasion. Whatever your opinions, there will always be times when, for the benefit of the horse, leg bandaging is necessary – for example, under the advice of a vet to bring down inflammation or when travelling in a trailer or lorry with another horse who may tread or trample on him when finding his balance. Some horses hate their legs being touched and have difficulty in accepting the length of time it takes to bandage legs as well as the claustrophobic feeling that bandaged legs can generate. If you are kneeling down attempting to bandage a leg, you are in a very vulnerable position if the horse chooses to kick out.

■ Causes

All horses should have their legs regularly handled as part of their general maintenance routine but all too often this is omitted from their general care. Horses are naturally fearful of having their legs lifted or restrained as always being ready to run is at the core of their survival instinct.

Previous injury or soreness (such as mud fever) can make horses highly sensitive to their legs being handled, long after the actual pain has gone.

The use of boots or bandages can itself cause soreness due to friction of the straps rubbing. The lack of air which causes the sweat that is often seen when the bandages are removed can encourage fungus and bacteria to breed.

Above: Handle the legs gently but firmly and stand to the side. A tentative touch may make the horse nervous but purposeful handling radiates confidence.

■ Remedy

Groom your horse's legs regularly with a **soft** brush. You can then see early signs of rubbed patches or mud fever which might cause problems later. Run your hands down the tendons on the back of each leg and down the front of the cannon bone and around the fetlock, pastern and coronet. This will also help you to identify any changes or swelling when diagnosing lameness.

Allow the horse to see and sniff any boots or bandages you are using. Enlist a helper to hold the horse still and soothe the horse. Handle the legs gently but firmly – a firm touch is more reassuring to the horse. Leg wraps are very quick and easy to use compared to traditional bandages and are ideal for protection during travelling, although if you require leg support, they may not be adequate for resting horses or over wound dressings – check with your vet.

If your horse insists on fidgeting then ask your assistant to lift one of his legs *(see below)* which will mean he has to stand firmly on the other three. This may be the best ploy to assist the task of traditional bandaging where one stamp by the

horse can undo all your hard work if the bandage has not yet been tied or taped to secure it. Your helper can feed the horse a titbit to make the whole thing a more pleasurable experience. The same method can be employed when putting on exercise or bell boots. Fortunately this job can be made simpler by choosing boots with Velcro fastenings.

Safety Warning
When bandaging or handling a horse's leg, always **stay to the side** and **crouch** rather than kneel on the ground so that you can move away quickly if a problem develops.

TAIL BANDAGES

▓ Problem

Some horses will clamp down their tail making it impossible to pass the bandage under the dock.

▓ Causes

It is just a natural reaction and most younger animals act in this way until they trust you. As horses age, their tail reaction seems to get much slacker as, just like everywhere else, their muscle tone deteriorates.

▓ Remedy

If you have a difficult horse and you try to haul the tail up you are likely to get booted out of the way! Life is made so much easier with the gentle approach. Stand to the side of the horse,

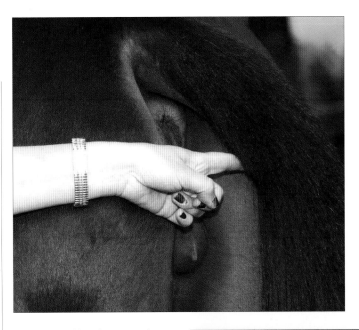

Above: By tickling the area under the dock where there is no hair, the horse will respond by conveniently raising his tail for you.
Right: This reaction allows you to apply the tail bandage safely to produce a neatly flattened tail.

tuck your finger in about 5cm (2in) down from the top of the dock and gently tickle the area where there is no hair. As if by magic, the horse will raise his tail. The bandage can then be passed underneath and this can be done as many times as needed. Cotton and neoprene tail sleeves are also available which negate the need to wrap the tail in the traditional manner but these do not lay the hair as neatly as bandaging does.

■ PULLING BACK WHEN TIED UP

■ Problem

You have a horse that either immediately rushes back when it is tied up, or which pulls and pulls against the restraint until the rope or headcollar breaks and he is loose. This can not only cause injury from the friction of the headcollar under strain but the horse may fall when the rope snaps. There is also a further danger to horses and people if the horse then rushes round the yard in panic when it breaks free. This problem is a nuisance for an owner as there are many times during the day when it would be helpful to tie a horse up while grooming or tacking up but when leaving him for just a second to collect the saddle will cause a problem.

■ Cause

Any form of restraint is difficult for an animal to deal with, but most horses soon learn that being tied up to be groomed or checked over is nothing to fear and remain calm. Some horses, however, have learned the pulling behaviour as an avoidance tactic and are just using their weight and strength as a way to avoid what they regard as an unpleasant experience or to get their own way. Others may have been fine in the past, but if something startles them while they are tied up and their natural flight response is disabled, they will scare themselves still further while trying to free themselves. This fear response may then be triggered each time they are tied up.

■ Remedy

You do not want a single event (for example pulling back when you have tried to hose a horse's legs) to escalate into a learned negative response to being tied up. Training to be tied up at a young age should be part of the normal handling process. The horse must learn to trust you and even allow you to lift one of his legs (which would further impede any chance of escape) and still remain calm. You will find that most horses will be far less likely to pull back if other horses that they feel comfortable with are tied up close to them.

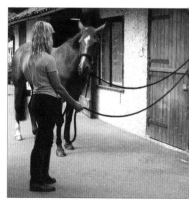

Above: You can be in control of the tension as the horse pulls back if you use a long rope that is threaded through a tie ring while training.

To teach the horse to accept being tied up, thread the rope through a tie ring on the wall (just for training purposes as this allows the rope to slide more easily than through a twine loop) but keep hold of the end instead of tying it. If the animal starts to back up, allow some slack keeping only minimal tension and encourage him to step forward again by pressure on his hind quarters until he is back in position and then say the command 'Stand'.

Above: This is a frustrating problem as the horse risks injury from the halter being under tension or it may break free and cause chaos in the yard.

Stable Vices and Management Problems

Progress to a lunge line threaded through the ring so that while the horse remains in position you are able to retreat further and he believes he is tied up alone (rather than attached to you). The next step is to tie him with a quick release knot to a breakable loop. Position a calm animal next to him and then move on to tying him

Left: The presence of a calm neighbouring horse can help to allay the anxieties of an animal that struggles when tied up.

up alone while you sit close by. A reward for patience may assist in reinforcing the idea that this is a pleasurable experience as will hanging up a haynet for him to pick at while tied up.

For difficult operations such as clipping or hosing, always have a helper holding the animal as even the calmest horses can become anxious.

Key points to remember are:

- Never tie directly to a fixed object – not even a tie ring. The exception is if you have a special release clip on the headcollar rope which will come undone under strain. In all other cases make a breakable loop from baler twine and tie onto that.
- Always use a quick release knot (see diagram).
- Do not leave too much rope slack which could catch on door bolts etc. and so frighten the horse.

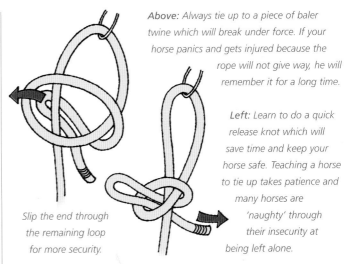

Above: Always tie up to a piece of baler twine which will break under force. If your horse panics and gets injured because the rope will not give way, he will remember it for a long time.

Left: Learn to do a quick release knot which will save time and keep your horse safe. Teaching a horse to tie up takes patience and many horses are 'naughty' through their insecurity at being left alone.

Slip the end through the remaining loop for more security.

- If you need a longer length of rope in order to feed from a bucket on the floor, do not leave the horse unattended.
- Never tie up a horse from his reins as, if he struggles, the bit may cause terrible damage to his mouth.

continued ➡

UNDOING DOOR BOLTS OR ROPE KNOTS

Safety Never use 'builders' buckets with handles attached, as these can become hooked on bridles and headcollars and cause even the calmest horse to panic with devastating effects.

Tip Generally horses are tied facing a wall. For a prey animal this is disconcerting as they will feel vulnerable not being able to see what is going on and who or what is approaching. They may become uneasy because of the clattering of wheelbarrows etc. behind them. It is also very boring for those left for longer periods of time. Consider constructing a hitching post or rail so that horses can face out and see what is going on.

UNDOING DOOR BOLTS OR ROPE KNOTS

Problem

While tied up the horse uses his muzzle to fiddle with the knot of the rope or door bolt and learns to undo it to set himself free. Obviously this is not ideal as the horse can then roam loose where he could suffer injury or cause damage.

Left: We rarely give our horses the respect and credit they deserve for showing intelligence. Fiddling with knots may be annoying but look at life from their point of view.

Above: Remove handles from buckets on the yard. Handles can get attached to headcollars when horses drink or eat and cause blind panic if the horse cannot free himself.

Cause

However irritating this must be for an owner, I am afraid I am of the school that cannot help but be impressed at this intelligent and dexterous behaviour. How can you blame an animal for doing this when its make-up and means of survival stem from being able to run from danger and graze while covering many miles each day, but who is now confined either by a rope or stable door? Of course, sadly we cannot in many instances grant them the freedom they desire but you must respect the intelligence of the horse for trying!

■ Remedy

Relieving boredom and hunger usually address the root cause. Feeding forage in small-holed haynets *(below)* – and not leaving the horse or pony for hours without food – plus giving the animal plenty of attention and exercise will help. These intelligent animals may well recognize the signs of any routine that pre-empts the chances of being turned out or ridden and this brings about agitated behaviour.

Even if you recognize the cause, you will still probably have to concentrate on preventing the action of undoing the knot or bolt as you cannot erase the intelligence behind it!

There are safety bolts on the market which prevent horses from moving bolts with their muzzle. To prevent ropes being undone, rather than just tying a more secure knot which may be too fiddly for you to undo later, it is best to make up a leadrope with a clip fixing at each end, which will be secure but very easy to use. It simply clips onto the loop of baler twine that attaches to the wall fixing and means that you do not have to tie a knot. Clip fixings can be bought at hardware shops or taken from another leadrope.

Above and left: Clipping a lead rope to the door bolt helps to prevent dexterous muzzles from undoing the door lock! Installing a lower kick bolt will also give added security and keep doors in alignment if they are repeatedly kicked by the horse.

FENCE BREAKING/ESCAPING FROM FIELDS

Problem

Most people strive to have smart, safe and secure fencing. However, some horses will manage to barge through or break many types of good quality fencing causing escapes and, more seriously, damage to other people's property or accidents on the road. You may then be faced with a bill to compensate for this damage plus any vet's bills and the cost of repairing the fences. These horses become a liability as the behaviour is like a ticking time bomb causing you to become anxious as to when it will happen again.

Above: A rail of timber is as fragile as a matchstick to a large horse. My intelligent Clydesdale X would not let anything stand in his path when it came to getting to new grass. Here the less bold Arab, not so keen to negotiate the obstacle, looks on as he moves upfield.

Causes

- Intelligent animals soon realize it is easy to use their weight to get through a fence.
- Separation anxiety can cause athletic animals to jump fences.
- Bored or lonely animal seeking stimulation or company.
- The grass is greener on the other side!
- Using the fence as a scratching post has weakened it.
- Poorly fenced field with gaps *(right)*, loose wire or wobbly fence posts.

Case History

I have a very clever Clydesdale X who is also known as Harry Houdini! Over a period of five years I have lost count of the times he has escaped from the fields I rented (often taking with him other horses and donkeys on a merry jaunt round the neighbours' gardens and even onto a golf course!). He and I were not popular and I soon became very proficient in lawn repair and needed a constant supply of presents to appease the irate neighbours.

Remedy

Now I have my own land which is Harry-proof, I can offer the following tips which have kept him in.

1 Dense hedgerow growing in front or behind a fence *(right)* is the most horse-friendly secure fencing. Trees such as blackthorn, hawthorn and holly fill the openings between the strands of wire or rails through which a horse might otherwise stretch his neck and help to stop him stamping on the wire.

2 Having hedgerow on the inside of the fence (or, if outside, left untrimmed to grow through it) also provides a beneficial additional food source plus branches to scratch against to relieve that annoying itching!

3 Rabbit netting nailed to post-and-rail fencing will stop horses reaching through with their heads.

4 If you prefer post and rail alone, position the rails close enough so that the horse cannot get his head between the rails. Raise the bottom rail high enough off the ground so that it is out of reach from being stamped on easily.

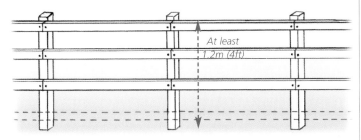

At least 1.2m (4ft)

Above: Keeping the rails close together and the bottom rung high off the floor will help to prevent damage from inquisitive heads and stamping hoofs.

5 Electric fencing is a godsend! It can either be used as an independent temporary fence within the existing boundary (useful for horses that jump fencing) or be attached with screw-in connectors directly onto your existing fence to act as a double barrier. There are very few horses who will rub against or try to reach over fencing once they have got a 'bolt' from an electric shock. Make sure it is positioned effectively.

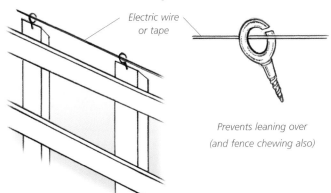

Electric wire or tape

Prevents leaning over (and fence chewing also)

6 Bear in mind that a horse with a rug on will not feel the sting of an electric fence unless it has one strand at nose height and another at knee/hock height.

7 Provide enough forage to keep the animals occupied. Provide branches and mineral licks in the field. Large logs and branches may not look very tidy but they will satisfy a horse's desire to chew bark. This aids mineral and fibre intake which may be lacking, even on good pasture. They will also act as scratching posts if you are not lucky enough to have a tree to provide this function as well as shelter.

8 In the wild, horses would naturally move over hundreds of miles and, similarly, domestic horses relish moving onto a new paddock. If you practise good field rotation, your horses will enjoy their changing vistas and good grazing. An overstocked, muck-ridden, weedy paddock is an invitation for your horse to look for greener pastures elsewhere.

Caution Never position wire strands so low that a horse can paw at them. I have seen horses with wire looped round their legs or caught between the hoof and the shoe, resulting in terrible injury if the horse struggles. The only safe wire to be situated low to the ground is rabbit or deer fencing where the mesh of the wire is too small for a hoof to pass through it.

Tip There are now electric energizers for electric fencing which come combined with a smaller integrated battery (lasting approx. 14 weeks). These are very useful for moving from field to field, rather than the traditional type which require a huge vehicle-type battery to be lugged about.

Right and left: Electric fencing is quite a cheap and effective fencing solution. It can also be used in conjunction with traditional fencing (left) to prevent wooden rails from being chewed or leaned on.

■ FENCE WALKING IN A FIELD

■ Problem

The horse is acting like a caged lion, pacing back and forth up the fence line, chopping a path in the mud/sand and generally looking miserable. You are not sure what to do – does he want to come in despite only having been turned out for an hour? Some horses even charge up and down the length of the field, skidding to a standstill which can be alarming to watch and they will not settle to graze the pasture.

■ Remedy

1 It is natural for horses to form close relationships with each other and to feel vulnerable when they are isolated. Even in a group, if a horse's pair-bond is taken away he is likely to feel anxious that they are parted even though there are other horses in the field. This behaviour usually subsides when the pair-bond returns. It is not usually necessary to shut the fretting horse in a stable and the behaviour often subsides over time when a sense of confidence grows that their friend will indeed return after his ride. It is helpful to ask someone to stay and keep an eye on the anxious horse and it will give you peace of mind if they tell you that the manic behaviour stopped within 15 minutes of you leaving the yard. Aromatherapy oils can help horses who feel fearful when

Above: Fence walking is a distressing sight – relentless pacing back and forth. Your horse should enjoy being turned out so why won't he settle and graze?

■ Causes

1 His companion has been taken away and he is anxious.
2 There is little grass in the field or he knows he is due for a feed when he comes in.
3 He is cold/wet/too hot and wants to come in for shelter. The flies are annoying him.
4 He sees people or other horses and wants to be with them.

Right: Flies can be a real annoyance and make horses unsettled.

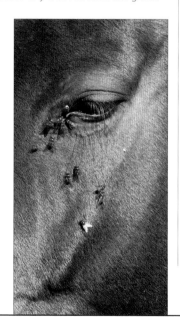

Above: Some horses thrive in a busy atmosphere and feel uneasy in a distant field away from the hub of activity. Try to locate them in a paddock where there will be lots of contact with people and other horses.

they are parted from their pair-bond. These can be offered to the horse to be inhaled, or can be massaged into the body. If preparing a massage, first take advice from a qualified aromatherapist regarding suitable oils to use and what degree of dilution is advisable.

2 So many times I hear people say 'My horse hates being out, he is always waiting to come back in.' It is very unnatural for horses to live in stables and almost always the only reason the horse wants to come back in is that there is no grass in the paddock and he knows that he will get a feed or haynet when he goes back to the stable. It is far healthier for a horse to self-exercise and have the freedom of the field but he desires food too. As an experiment, put the feed bowl and the hay ration in the field before turning him out. When he learns that this is where food is to be found, he will not choose the stable! A manege or yard area can also be used and the haynets hung round the edge (or arranged in piles on the floor). By scattering carrots or pony nuts across a wide area, it allows the horse to satisfy his foraging behaviour rather than wolfing his food in one go and then having nothing to enjoy.

3 Horses have a fine natural system for regulating temperature as their coat will flatten or lift to trap air which provides an insulating effect. It is rare for horses and ponies (other than elderly or sick animals) to get uncomfortably cold in dry frosty conditions, but when wet this function is diminished and horses can soon get a chill or even rain scald so a waterproof rug may be the answer. Heat and sunlight can upset horses so shelter should always be provided. If you have felt the reduced surface temperature of a grey horse who has been in the sun compared to a bay or black animal, you will appreciate the benefit of using a light-coloured cotton sheet on a dark-coloured animal to help reflect the Sun's rays. Horses also need protection from biting insects, another reason many horses seek shelter in the stable. Using fly repellent (especially on the face and belly areas where the skin is thin) will greatly alleviate irritation. Alternatively turn the animal out overnight when the heat and flies are far less active.

Above: Stabled horses get their rations in bulk whereas in the field they have to 'work' for it. Scatter the provisions of hay in the paddock, manege or corral, so that they get the best of both worlds.

4 Some horses are naturally gregarious and want to be at the centre of the action. They don't want to stand in the field alone when there are horses and people milling around in the yard. If you feel that your horse would appreciate more attention, then

consider a sharer. More work and stimulation will improve your horse's mental and physical well-being as well as lessening the guilt you may feel when you are rushing around feeding and mucking out but barely have adequate time to give your horse the love he deserves.

Left: In hot temperatures many horses seek the solace of their stable. A fly mask and field shelter that provides shade will make them far more comfortable outside.

■ FEEDING NATURALLY = BETTER BEHAVIOUR

Left: Nature offers a vast larder of foods to enhance your horse's diet.

while acid will build up on account of the long periods of time that are spent with an empty stomach.

Even performance horses should be treated like horses rather than machines and not forced into unnatural feeding regimes. Trickle feeding of good quality forage can provide almost all the energy requirements needed by even the hardest-working horses and this can just be topped up with small feeds as required. Feeding naturally will **enhance** performance and help prevent stereotypical behaviour brought on by stress, like weaving and crib biting, as well as limiting digestive complaints such as colic and ulcers.

A horse should be fed as nature intended and this means providing a diet made up of predominantly fibre that is eaten throughout the day and night. This contrasts starkly with the 'three meals a day' hay and feed given in many stables (or more usually only two meals given at around 8am and 5pm) which usually consists of a large bowl of feed and a haynet which will only last a couple of hours. Feeding a horse in this way causes him to bolt down the food the second it arrives and then stuff himself with the hay straight after. This often overfills the stomach giving the nutrients in the precious hard feed little time to be absorbed before they pass into the hind gut. Blood sugar levels will also go through a series of peaks and troughs

A horse's metabolic make-up is designed to allow it to graze almost constantly but this feeding pattern is very difficult to achieve with stable-kept horses who may become obese if offered a constant supply of hay. Small-holed haynets can be used as well as using different forages to satisfy the grazing motion without piling in the energy. Thrashed hay, short chops and chaffs will all add interest to the diet and I use barley straw for bedding and positively encourage my horses to eat it! Many problems involving horses bolting food or suffering choke are usually caused by the horse trying to ingest it too quickly after a long break from eating. Trickle feeding forage takes away this desperation.

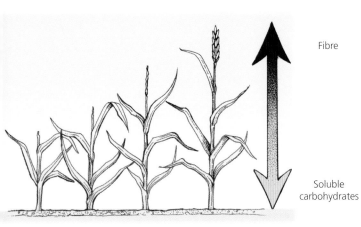

Above: Long does not always mean lush! Regularly topped or young leafy grass is actually higher in soluble carbohydrates and lower in fibre than grass allowed to go to seed which has a higher fibre-to-carbohydrate ratio.

Above: This snack of cleavers and rosebay willowherb is far more beneficial to a horse's well-being than just adding a scoop of vitamin supplement to his feed.

Field management is therefore highly important and the grazing has to be suitable for the horses eating from it. The growth of grass can be controlled by sharing the grazing with other horses or sheep or by deciding whether or not to fertilize the land.

In the wild horses would naturally cover 10-20 miles (16-32km) a day seeking good vegetation and water sources while keeping fit and burning energy at the same time. Mares might well have the added burden of foals at foot and any extra weight gained during months of bountiful vegetation would help to carry them through the shortages of harder times. The risk in a domestic situation is that the paddocks we offer are often tiny in comparison and the energy expended by horses is minimal during grazing hours and negligible when stabled.

Field rotation systems are needed in most cases to balance out the need to provide adequate grass for the work and type of horse and to prevent fields from becoming overgrazed and stale. Temporary fencing can be used to deal with fluctuations in grass growth and to separate horses into different areas.

Branches (try willow, hawthorn and fruit trees) and leafy brambles are excellent roughage and should be put in both the field and the stable. Horses will strip bark and pull the black-berry leaves from the thorny stems and benefit from the vitamins contained in both. Rosehips are another free treat and contain excellent levels of both Biotin and Vitamin C. Nature offers us a salad of delicacies and I regularly pick plants and herbs when I am out walking my dogs to feed to my horses later.

Succulents like carrots, parsnips and swedes can be fed in large quantities to provide a filling and nutritious snack without introducing high calorie levels at the same time. One comment though – do **NOT** cut up the carrots – most horses have strong, sharp teeth and derive pleasure from using them to nibble at even the largest, hardest root vegetable. By allowing your horse to express natural foraging behaviour in this way, he will be far happier.

Above: Succulents like carrots are a treat that even fatties can have in quite a high volume.

■ METHODS OF CONTROL

I really cannot emphasize too strongly how sympathetic handling and time spent patiently helping your horse to overcome his fears is the route every owner should strive to follow for the long term well-being of the horse. To be honest, reliable 'quick fixes' do not exist, but I am the first to admit that, however much you know, it is hard always to have sufficient time and resources available to devote to the horse.

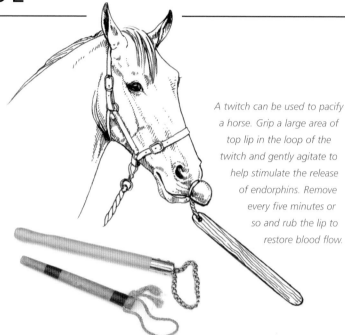

A twitch can be used to pacify a horse. Grip a large area of top lip in the loop of the twitch and gently agitate to help stimulate the release of endorphins. Remove every five minutes or so and rub the lip to restore blood flow.

However, there are bound to be occasions when we do need instant results. We may wish to have the horse shod or perhaps clipped in time for a particular event and several days or weeks of acclimatization lessons just don't fit into our schedule. We need it done now!

Similarly there are horses who have such deep-seated phobias that they are virtually impossible to overcome and they may have exhausted the patience of the handlers attempting to understand them. There are also instances where the safety of the horse and/or handler would be compromised by

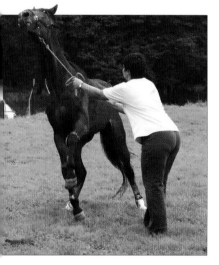

Above: Their natural response to danger is 'Fight or Flight' and so handling fearful horses can make life hard for a horse owner.

struggling to work with an overly anxious horse. In these instances it is best to just try and get the job done quickly, safely and with the least amount of stress and that is when one of the following methods may be employed.

THE TWITCH

The practice of twitching has been employed all over the world as an effective method of restraining a horse. The twitch itself can either be a proprietary shop-bought metal 'nutcracker' type of instrument where the horse's upper lip is gripped between

the two handles, or a home-made 'rope noose' type where a loop of cord attached to a handle is twisted to grip the top lip in the loop.

Whichever type of twitch is used, it is important that as large an area of top lip as possible is enveloped in the twitch and the tension is repeatedly loosened and tightened. If you consider that clipping a horse may take 40 minutes or more, restricting the blood flow to the extremity of the lip for this length of time is not a good idea. Therefore it is essential every five minutes or so to remove the twitch and rub the lip area to restore the blood supply to it, before replacing it. When applied, the twitch should be gently agitated or rocked to stimulate the release of endorphins (the body's pain-killers). The stimulation of the sensory receptors in the mouth transmits signals to the brain allowing pain-relieving endorphins to be released into the body.

It is not clear if these endorphins are produced simply as a response to actual pain experienced or if, more likely, the twitch has the same effect as an acupuncture needle, where stimulation of a target area of apparent pain (the needle point) produces a far greater overall sedative effect on the body and a sense of well-being. The production of endorphins is a normal part of daily life; they are associated not only with fighting pain but also are released during eating, social interaction (e.g. grooming) and sexual behaviour, as well as during typically addictive or repetitive behaviour, such as crib biting seen in horses.

As an experiment, if you grip and agitate your own top lip you can actually begin to feel slightly light-headed. Some people also claim to also achieve good results by clasping a fold of the horse's neck skin in their hand, although trying to test this on myself certainly has not had the desired effect!

Above: Do not sedate a horse that is already distressed. Leave the job for another day. Use the sedative before you start, while the horse is still calm.

LOCAL ANAESTHETIC

For tail pulling or when a horse needs an injection, for example, it is possible to use a topical application of an anaesthetic cream to dull the senses in that particular area. This local anaesthetic is an excellent way of allowing you to perform an action that would otherwise be painful and stressful to the horse. However, you should be aware that, like other drugs, it can be detected in both urine and drug tests for several days, and so what may seem to be an innocuous treatment on a tiny area of the horse's body may subsequently cause disqualification if you are tested after a competition.

SEDATIVES

Equine sedatives are prescription-only medicines – they have to have been prescribed by a vet for the one particular horse they are intended for. There is good reason for this. Such drugs should never be used indiscriminately due to the sensitivity of the dosage and the potential side effects of these substances which work on the central nervous system.

There are two types offering light or deep sedation. A common light sedative is acepromazine (ACP) which can be given orally as a gel, paste or in tablet form *(below)* – although the tablets are more commonly used on small animals. It can also be given by injection by a vet. Even a vet can find it tricky to quantify the dosage for a particular animal especially as the drug can take up to 45 minutes to take effect. It is essential to allow plenty of time both before and after administering the drug (as its effects can still be obvious

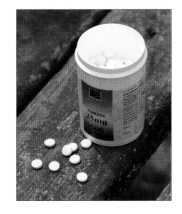

five or six hours later). A light dosage might be barely noticeable and can be used just to 'take the edge' off a nervous or lively animal, for manipulation of limbs or loading into a lorry, where it is important for the horse to maintain his faculties.

Injected sedatives work more quickly and alpha-2-agonist drugs produce far deeper and more instant effects where a horse can become quite oblivious to his surroundings.

CHAPTER THREE

EQUITATION AND SCHOOLING PROBLEMS

Being able to control such an amazing animal as a horse is something that I still find incredible. The fact that by subtle aids we can communicate with a huge and powerful animal that is willing to carry us and work with us both in competition, as well as in times of leisure and companionship, is extraordinary. Horses aren't 'push-button' machines and we must accept that we will encounter problems achieving our goals. It helps to appreciate just how accepting your horse is of you and to look at your demands from his point of view.

■ HORSE RESISTS BEING BRIDLED

■ Problem

It sounds simple to put a bridle on a horse, but I have to admit it is no mean feat to coordinate the positioning of a bit attached to a jumble of leather straps onto a horse's head. Firstly, you risk knocking the horse's teeth with an unforgiving lump of metal (the bit) and risk scraping a buckle over the horse's eye attempting to stand on tiptoe to reach up to pass the browband over first one and then the other ear of a tall horse. Squashing the ears and catching tufts of forelock in the wrong place resulting in a 'bad hair day' for the horse are other common problems! You can expect such problems even with a cooperative animal and unfortunately these problems are only exacerbated if the horse chooses to poke his nose skyward or clamp his mouth shut.

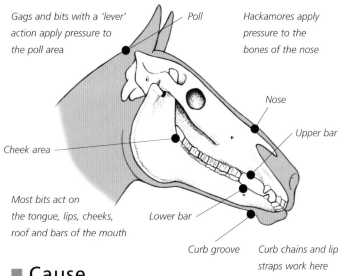

Gags and bits with a 'lever' action apply pressure to the poll area

Poll

Hackamores apply pressure to the bones of the nose

Nose

Upper bar

Cheek area

Most bits act on the tongue, lips, cheeks, roof and bars of the mouth

Lower bar

Curb groove

Curb chains and lip straps work here

■ Cause

The simple logistics of a small person bridling a tall or uncooperative horse make the exercise difficult. Added to this, the horse may well not like his bit, or the way you knock it against his teeth in an effort to force him to open his mouth, or the overall restraint that he feels when wearing a bridle. It is rare to see horses be so evasive when a headcollar or halter is put on so it is likely that it is the mechanics of the bridle or the clumsiness of the handling that are creating the problem.

PARTS OF THE BRIDLE

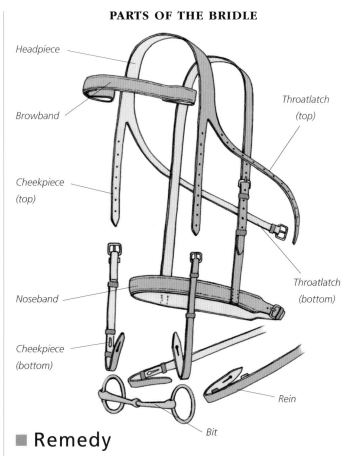

Headpiece

Browband

Throatlatch (top)

Cheekpiece (top)

Noseband

Throatlatch (bottom)

Cheekpiece (bottom)

Rein

Bit

■ Remedy

To avoid being rough with the horse you need to be able to reach the top of his ears easily so that positioning the leatherwork can be done accurately without any fumbling. Standing on a box will gain you the necessary height but this is only practical if your horse remains still and does not back away from the box. In any case, you will not always have a box to hand so you should attempt to find a more permanent solution.

Teaching Your Horse To Lower His Head
There is normally less resistance to putting on a halter or headcollar and by using this as a tool, or merely by employing

hand pressure, we can teach a horse to lower his head on command. Horses naturally lean into pressure so they do have to be taught to move away from pressure.

1 Have your horse in a halter or headcollar in a quiet environment so that his full attention is on you.

2 Simply pulling down hard on the rope of the headcollar is unlikely to have any beneficial effect at this stage and the horse may well throw his head up in protest.

3 With the rope in your left hand, reach up and place your right hand just behind his ears at the poll area. Stand on something like a sturdy box or crate if you are likely to lose the position if he tosses his head around.

4 Very gently apply downward pressure with the thumb and two fingers of

your right hand while holding the rope in your left hand. Try to keep your hand in place on the mane even if your horse resists.

5 Say 'Lower' as you apply the pressure.

6 If there is no reaction, increase the pressure. You may have to pinch the poll area between thumb and fingers of your right hand.

7 As soon as your horse lowers his head even slightly, release all pressure.

continued ➡

Left and sequence below: In order to teach your horse to lower his head when asked to do so, apply steady downwards pressure behind the ears and draw down on the leadrope. Say 'Lower' and as soon as the horse yields, release the pressure but hold his position. Do not try to keep pushing further and further all in one go. Ask again in small stages for more lowering and praise him at each stage.

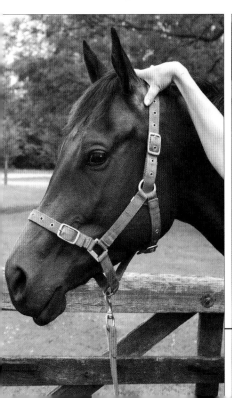

■ HORSE RESISTS BEING BRIDLED

8 A common mistake is to keep pushing the horse further and further down in one go. Press down then allow the horse to lower to avoid the contact but do not continue to grip as he goes down.

9 Imagine your right hand is on a static line. Your hand will remain at that set level and the horse will come away from it.

10 When the horse lowers his head stroke him or give him a treat.

11 Repeat the sequence:

- Give the command 'Lower'.
- Maintain downwards contact with your left hand on the rope.
- Start with a light touch on the poll area.
- Increase this downward (or pinching) pressure.

Below: This shows just how far you can lower the head. Horses soon learn what is required of them and this will make life easier for the pair of you.

Above left: It is far easier to bridle a compliant horse and in turn this allows you to be more gentle as you pass the bridle past the eyes and over the ears.
Above right: By being consistent in your approach and using a vocal command such as 'Lower', your horse will better understand what is expected.

- When the horse yields to the pressure, immediately release.
- Stroke and treat the horse.

12 You should be able to gradually increase the time of the downward pressure to ask the horse to lower still further.

13 Practise with the horse to help you when you need to put on martingales, neck straps etc. plus, of course, when you want to put on and take off the bridle itself.

14 By repeating this exercise the horse will soon anticipate your  wishes and very little downward pressure from a halter or your fingertips will be required for the desired result to be achieved. You may soon only have to say the word 'Lower'.

Left: A bitless bridle like this would help horses who resent anything in their mouths, but sadly these are rarely allowed in competitions.

Above: Most English bridles are overly restrictive and must feel claustrophobic for the horse to wear.

Now we have got the horse to play his part, it is up to you to act sensitively too. If, for example, someone wanted to apply make-up to your face or clean your teeth with a toothbrush, I am sure you would feel vulnerable while this was being done, particularly if in the past you had been bruised by one of the implements. The face is such a sensitive area and so it is essential that you take pains to be as gentle as you possibly can when bridling your horse.

Do not thump the bit against the incisors but cup it in your left hand in position against the horse's lips or teeth as you ask your horse to open its mouth by gently poking the thumb of the same hand in the gap between the horse's front and back

Above: Many western bridles use the minimum of leather and are therefore light and comfortable. Compare this with the English bridle shown above left.

teeth (where the bit usually lies). When passing the leatherwork over the ears, pay attention to the horse's eyes. With the head lowered there is no excuse for the bridle to be left dangling over one ear as you fish around trying to position everything blindly on the other side.

Teamwork makes the job far easier.

■ CROSSES JAW/STICKS TONGUE OUT

■ Problem

This problem is more commonly seen in classical English riding than Western riding as the contact from the rider's hands through the reins, bit and the horse's mouth is maintained throughout. The horse tries to evade the action of the bit by throwing his head about, twisting his jaws or going along with his tongue hanging out. The result of this is any 'outline' is lost and the line of communication from the rider's hand to the horse's mouth is broken which means that he cannot be controlled properly.

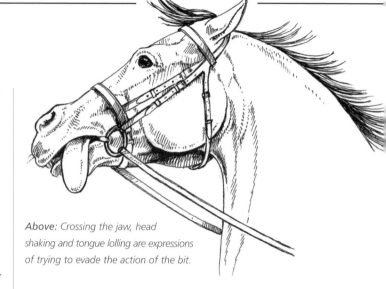

Above: Crossing the jaw, head shaking and tongue lolling are expressions of trying to evade the action of the bit.

■ Causes

All youngsters and newly bitted animals will play and mouth any bit (and, indeed, chew the reins etc. if they manage to get hold

Above: Some horses fiddle with the bit as soon as bridled while others resist when a contact is taken.

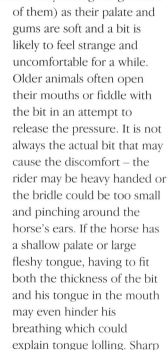

of them) as their palate and gums are soft and a bit is likely to feel strange and uncomfortable for a while. Older animals often open their mouths or fiddle with the bit in an attempt to release the pressure. It is not always the actual bit that may cause the discomfort – the rider may be heavy handed or the bridle could be too small and pinching around the horse's ears. If the horse has a shallow palate or large fleshy tongue, having to fit both the thickness of the bit and his tongue in the mouth may even hinder his breathing which could explain tongue lolling. Sharp teeth are a common problem and they can ulcerate the gums which can cause tremendous pain when the sore spots are in contact with a bit.

■ Remedy

I am utterly sick of seeing every English bridle sold with a flash noseband attached and almost every horse wearing some kind of tie-down noseband. **A good horse:rider partnership should not be based on restrictions, but on freedom.**

Firstly, take a close look at the conformation of your horse's mouth and take into account the size and mechanics of this when choosing the thickness of the bit. Have his teeth checked and, if necessary, any rasped at least once a year. Consider whether he usually goes better in a straight bar or jointed bit. If you attach a straight bar bit and a jointed mouthpiece onto the same bridle, the jointed bit will hang lower in the mouth than the straight

Above: Many bridles seem to come with a tie-down as standard. I would rather find out if my horse is tense and uncomfortable!

Equitation and Schooling Problems

Above left and right: The canter starts well (right) but the horse begins to set his jaw against the rider's contact. The martingale has the negative effect of making the contact still more rigid (left).

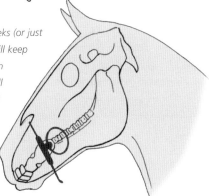

Left: A full cheek bit such as a Fulmer snaffle has a lateral guiding effect and can also have mullen, double-jointed, waterford, slow twist, corkscrew, single or double-wire mouthpieces.

Right: A bit with full cheeks (or just upper or lower cheeks) will keep the bit centrally in position whereas an eggbutt, small ringed or even a D-ringed bit could be pulled through an open mouth by a rider trying to control an awkward horse.

bar so bear this in mind when adjusting the height of the bit by the buckles on the cheek pieces of the bridle.

Ride sensitively using the lightest of touch and use a neck strap if there is the slightest chance you will become unbalanced in the saddle and hang onto his mouth.

It can be very dangerous if the horse opens his mouth and there is a risk that the bit might slide through into his mouth. You can avoid this by using a full cheek Fulmer snaffle or similar high cheek or long-shanked bit as this will keep the bit central in the horse's mouth.

Any type of tie-down noseband should be a last resort as I would rather see when my horse is tense and unhappy, rather than force his mouth shut and so cause tension in the jaw and a horse that resents bridling.

Goal

In an enclosed area tie your reins in a loop to gather most of the slack and ride just using your leg and seat aids and voice commands. If you need to check the speed or steering use only one finger if possible to adjust the reins. We so easily become reliant on the reins which should really be for fine tuning your aids rather than as a means of rough restraint.

■ DIFFICULT TO MOUNT

■ Problem

The problem of having a horse or pony who fidgets, moves away or walks round you when you are trying to mount is quite a common one. What makes it dangerous, rather than just irritating, is that you could be standing on one leg with the other foot tucked into the stirrup iron and therefore easily unbalanced or even dragged along if the horse moves off. Some horses allow you to get on, only to rush off as soon as your backside hits the saddle, which can be very unnerving if you haven't been able yet to position your feet securely in the stirrup irons.

■ Remedy

By changing the normal mounting routine, you will get the horse to start listening to you and so enable you to retrain him. The easiest way to do this is to change the location of where you mount. Face the horse away from the yard exit, either into a corner of the stable yard or towards a fence, to limit his ability to walk off. If you have a static mounting block in an open location, ask a friend to stand and block your horse from moving.

Consistency is the key. Lead your horse into position and ensure he is standing squarely. If not, he is likely to move when you mount in order to balance himself under your weight. Use a

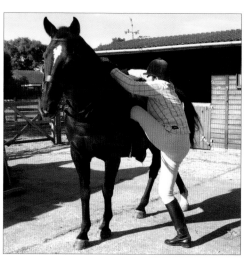 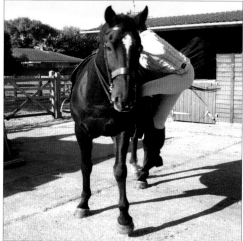 

Above left to right: It is frustrating and unsafe to have a horse that fidgets when you are trying to mount. With your foot in the stirrup iron you are vulnerable to being dragged. Also, when showing, you may be required to mount while your horse stands quietly, so you risk being penalised if it misbehaves.

■ Cause

Sometimes there may be an underlying fear associated with discomfort of the saddle or the weight of the rider (see Cold Backed, page 108). If this has been eliminated, then it is often the case that the horse has not been correctly disciplined to stand each and every time that you want to mount, and it is simply displaying over-keenness to move off.

mounting block which does not wobble or rattle which may unsettle the horse. Hold the reins evenly so the horse's neck is not turned to one side which would unbalance him, nor should the reins be shortened too much which would raise his neck and hollow his back just when he needs to lift it to take your weight. Use a recognizable command such as 'Stand' or 'Halt' and attempt to mount. If he moves even one step, reposition him and begin again. A helper is a good idea to prevent you from getting cross and frustrated through having to move on and off the mounting block!

If he does stand quietly, do not rush to get on but instead make a conscious effort to pass your leg over his back carefully

Equitation and Schooling Problems

  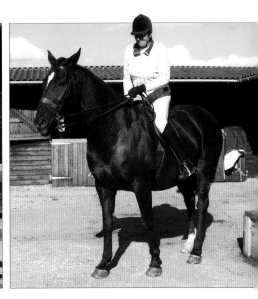

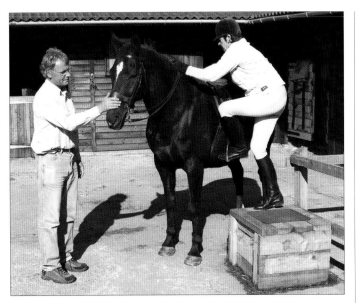

Above: An assistant will make the re-learning process easier. He can steady the horse and reward good behaviour. Start with actual restraint and then get the assistant to stand at a distance and only intervene when necessary.

Above left to right: Using a block makes life easier for both of you. It is less likely that you will unbalance your horse by hauling on one side of the saddle and easier to lower your weight down gently. Moving off can just be a habit that needs re-learning correctly. Insist that he stands, even if it takes ten attempts.

and lower your weight into the saddle gently – probably something you have not done in a long time!

Finally, **sit up**, **smile** and **stroke** your horse. It may help to offer your horse a titbit once you are safely in the saddle (or get your helper to give it) as a horse that is hopeful that a treat will be forthcoming will be less likely to want to charge off. Horses are intelligent animals and will soon learn that standing still saves a lot of time and argument and achieves the reward of a treat or of being taken out and exercised.

Important: If other people ride your horse, it is important that they maintain the consistent approach that you have established.

Safety tip: Never mount inside a stable or anywhere with low beams or doorways. You will be sitting tall on the horse's back and run the risk of hitting your head.

■ COLD BACKED

■ Problem

This term refers to the flinching behaviour of a horse when a saddle is placed on his back. He may sink or arch his back when being saddled or mounted and in some cases can pull back or try to sit down to avoid being mounted.

■ Causes

A young or unfit horse maybe 'soft' and lacking in muscle. It is understandable that it will be a slow process to strengthen and desensitize the area under the saddle. If an animal has experienced bruising or back pain in the past, it may be a learned habitual response. Short-haired, thin-skinned animals are more predisposed to friction soreness.

Below: Even something as simple as rolling or bucking can cause muscular strain and compensating for this pain can result in further stiffness. Muscles need to be strong and elastic to achieve easy motion. Muscle spasms may be a cause of a horse flinching when he is saddled up.

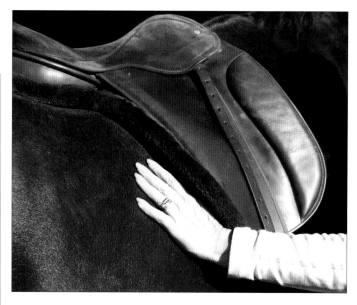

Above: A soft numnah will offer greater comfort by reducing pressure points and absorbing sweat which will lessen the friction that develops between a sweaty or wet horse and the leather of the saddle.

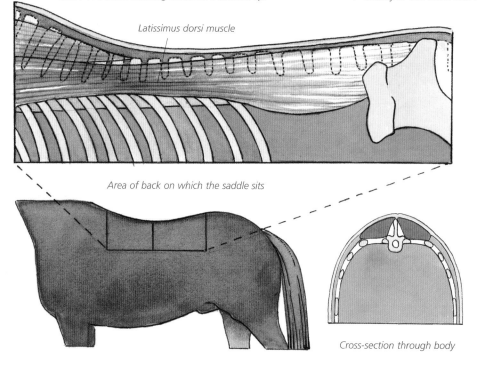

Latissimus dorsi muscle

Area of back on which the saddle sits

Cross-section through body

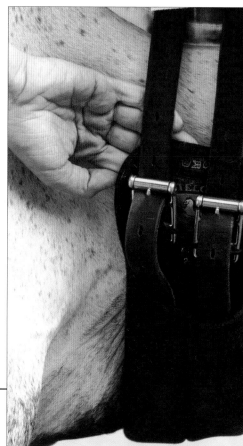

■ Remedy

If you encounter either of the above problems, firstly every effort should be made to have the horse's back checked by a qualified practitioner to rule out genuine injury or soreness.

- Have the fit of your saddle checked by a qualified saddle fitter, especially if you notice areas of the coat which have been rubbed or any lumps or sores that have developed.
- Use a soft cushioning numnah and ensure it is free from grease or mud which could chafe against the skin.
- Check the fit of the girth – is it too wide? Does it pinch? Feel for girth galls which could be caused by pinching or rubbing, especially when sweaty.
- Always tighten the girth slowly and gently in several stages.
- After tacking up, walk the horse in hand or lunge for a few minutes to warm up and 'bed in' the tack and then readjust the girth if necessary.
- Always mount off a mounting block to minimize saddle slippage and any tendency to unbalance the horse.

Above: Prior to mounting, walk the horse in hand and turn him in tight circles in both directions. This should show up any discomfort with the tack. Lungeing is a good way to 'bed in' the saddle and warm up the horse before riding.
Below: Tighten the girth in stages. I find the best girths are those with elastic inserts which give a little and so are kinder to the horse.

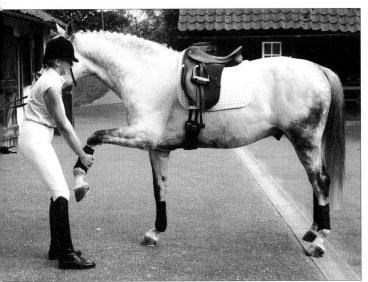

Left and above: Smooth out the wrinkles from under the girth which could pinch uncomfortably and ensure the girth is clean. Stretching out each foreleg in turn will lay the skin flat as well as being a good exercise for the horse.

■ BUCKING

■ Problem

This problem needs little description – when you have been badly bucked, you generally know about it! The horse's head goes down and his back end comes up, sometimes just once before a canter, but there are some horses who act like they are in the rodeo ring. I know from personal experience that after having been put on the ground several times, it really starts to knock your confidence.

Left: Very few people make their career or enjoy being bucked off like this rodeo rider. Most of us like to feel secure in the saddle.

■ Causes

- General exuberance, release of pent-up energy.
- Sore back – weak muscle tone or badly fitting tack.
- Immature or unbalanced individual.
- Confusion or stress – perhaps you are progressing too quickly with schooling or the horse is panicking.
- Insect bite, external irritation.

■ Remedy

Bucking is a natural expression of well-being and happiness in all equines. It is lovely to see horses prancing, bucking and snorting – but ideally only when loose in the

Right As well as anger or irritation, horses buck as an expression of excitement and happiness. 'Horse-watching' will make you a good judge of their moods.

field! A horse that bucks under saddle just once due to the anticipation of a good canter along with his friends is understandable. It is a form of self-expression and should be ignored, provided it is not violently unseating.

If you have a fit horse who is stabled, it is also to be expected that he will be like a coiled spring once given the chance to move. More turnout or lungeing before riding will help take the explosive edge off his behaviour and allow him to focus on your commands once this 'ping' is out of his system.

Any pain is likely to cause resistance, so regularly check your horse's back for any rubbed hair *(right)* or lumps caused by pressure points and remember your saddle may need re-fitting as your horse changes shape due to weight gain/loss or changes in fitness levels. Also check that the girth is not pinching the skin and the hair is lying flat behind it. It is helpful to lead your horse or pony round in a tight circle in both directions to iron out creases and allow the saddle to warm up before you mount.

Young horses or those brought back into work after a lay off will have diminished muscle tone and cannot be expected to work for long stints. Their workload needs to be increased gradually to harden the tissues supporting the weight of saddle and rider and to allow the horse time to learn to rebalance with a rider on board. A horse that is stiff or tense will benefit from massage, 'carrot

Above left and right: Release tension in the back and improve muscle tone by doing 'carrot stretches'. Encourage him to bend round to either side. For a forward stretch tempt him to reach for a carrot down between his forelegs.

stretches' and working long and low in an active walk to strengthen his back.

Often we do not fully appreciate how much harder it is for the horse to work with a rider on his back. It is difficult for an immature or unfit horse to turn quickly and attempting to gallop with an unbalanced horse can sometimes elicit a bucking fit. Often the horse wants to do what we are asking but it is just too much, mentally or physically. If this is the case, step back a couple of stages with his training to maintain his confidence and do not introduce anything further until you know he is really ready for more.

Irritation from an insect bite or a stone flicking up can cause a bucking reaction or the horse could simply be having an off day.

Riding A Buck

In order to buck the horse lowers his head, arches his back and kicks out his hind legs. If he cannot put his head down he cannot buck, although he might feel 'humpy' under the saddle. If you feel he is likely to buck, keep your reins very short, encourage him forward with your legs and voice and turn him in a tight circle. If he does catch you out, there are two methods of staying on you can employ.

1 Ride with a knot in your reins (to shorten them) or bridge your reins. If you get thrown forward you can then keep your arms straight, stand in the stirrups and lean on this bridge to keep you on board. Give the horse a smack on the neck or shoulder (never behind the leg) as a reprimand and to draw his attention back to you.

2 Think upright, upright, upright and keep your eyes up and your back straight and try and sit to the bucks. You may well bounce about and you may lose your stirrups but you should end up still with your bum in the saddle! At the earliest opportunity smack him on the neck with the whip and try to unbalance him by pulling on one rein only to turn his neck to one side to prevent further bucks.

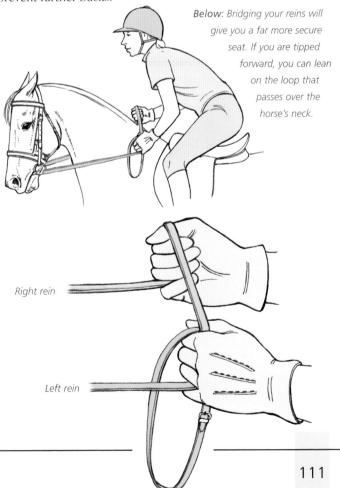

Below: Bridging your reins will give you a far more secure seat. If you are tipped forward, you can lean on the loop that passes over the horse's neck.

Right rein

Left rein

■ REARING

■ Problem

Rearing can be a problem in youngstock when they are being handled from the ground. It is very dangerous for a handler to have to contend with hooves at head height so these animals need authoritative education. A ridden animal that is known to rear is most undesirable and its saleable value if it has such a problem will be dramatically reduced.

Left: If you observe horses in a group you will see them challenging one another, rearing up and striking out with a foreleg. This behaviour can also be seen if they encounter a strange object or an animal that has taken them by surprise.

Below: If your horse rears, he is trying to escape from something. He is either evading your riding or something you are pushing him towards. Try to see the situation from his point of view.

■ Causes

Young animals rearing are displaying naturally playful and competitive behaviour. Rearing is predominantly prevalent in young colts and stallions. It is also a natural reaction to threat providing a means of attacking something that the animal fears (i.e. it is becoming as large and aggressive as possible when faced by a predator).

Ridden animals often rear as a form of napping or avoidance behaviour.

■ Remedy

An overly playful or dominant animal should only be handled by assertive people who can command obedience from the horse. Chiffney bits are well known as the 'anti-rear bit' and many people use one in connection with a bridle headpiece in place of a halter. They have a severe action if the horse misbehaves by putting pressure on the tongue so should only be used in responsible hands. A young animal may well grow out of this behaviour and castration will curb the natural urge towards competition in young males.

It is not uncommon for horses to rear at the most inopportune moments, such as when entering a show ring to compete. This tends to be a reaction stimulated by fear or an expression of protest – he is trying to escape but is prevented

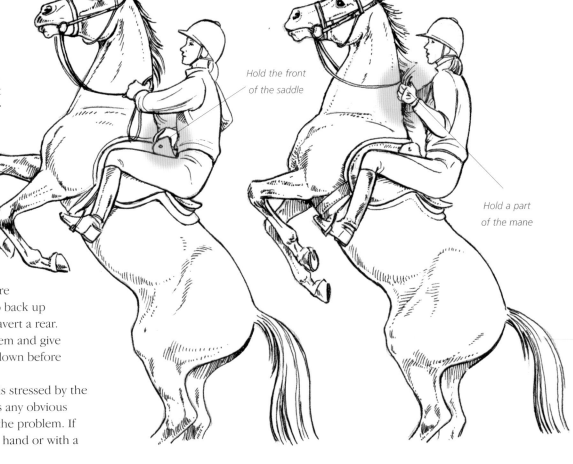

Hold the front
of the saddle

Hold a part
of the mane

from fleeing and so the only available direction is up. It can also be experienced as a direct response to rough handling or pulling on the bit so being a sensitive rider is paramount.

If a horse does rear up when you are riding, lean forward and do not hang on the reins. Grab a section of mane or the pommel of the saddle and lean forward. Releasing pressure and turning or allowing him to back up when you feel resistance may avert a rear. Walk him away from the problem and give yourselves both time to calm down before facing the obstacle again.

It is likely that the animal is stressed by the situation so establish if there is any obvious pattern in the occurrences of the problem. If there is, educate the animal in hand or with a

Above: Try not to hang on the reins or the horse could go over backwards. Release the rein pressure and keep your balance by holding a chunk of the mane or the front of the saddle.
Left: A common problem is napping and rearing as you attempt to enter a show ring. Practise on a schooling day or in a 'mock' show environment when you are calm and need not rush.

schoolmaster companion to increase his confidence. Practise a show ring environment on a 'schooling day' or mock up a situation where you can take your time and there is not so much pressure on horse or rider, and so accustom your horse to responding calmly to the experience that is unnerving him.

■ BOLTING/RUNNING AWAY WITH YOU

■ Problem

Unfortunately there are numerous instances where horses have run away with their riders who are unable to limit the pace. The term 'bolting' also refers to a horse running away in an effort to flee something it genuinely fears. The end result is that the rider is out of control and therefore likely to come upon hazards like ditches, pot holes or even road traffic at full speed and without due caution. The hazards can potentially be fatal.

Above: Galloping out of control is hazardous as everything happens so fast. Uneven ground or ditches can cause a fall if your horse is blindly charging.

■ Causes

- The horse may have been spooked by a sight, smell or sound and genuinely be trying to run away from the source of its anxiety. Even a branch caught in his tail can give the sensation that the horse is being chased by something and impart a flight reaction.

- There could be genuine pain, e.g. mouth soreness or a burr under the saddle, and he is trying to evade this aggravation.
- A horse with pent-up energy is very likely to turn a controlled canter into a full gallop and equally likely to use an 'excuse' (such as a plastic bag flapping in the hedge) to take off with you on board.
- A horse who has previously raced may view any section of grass as a cue to set off at high speed.

■ Remedy

A relaxed rider will help to keep a horse calm whereas a nervous rider is likely to heighten his adrenalin. Always look ahead for hazards that may alarm the horse and ride out in the company of safe and stoic horses. If yours is the type of horse who is actually bombproof but who finds excuses to misbehave, then keep a short contact and ride assertively. However, do not suddenly pick up the reins of a skittish animal or they will think that there is something to be nervous about.

If you are unfortunate enough to get bolted, try not to 'jump ship' as ideally you want to attempt to bring the situation under control. Keep your balance by holding a section of mane or neckstrap or bridging your reins. Sitting down on some horses seems to give them extra propulsion whereas others can have their stride stunted by the rider's weight, so you will have to judge whether or not to lean forward or sit to the stride.

If you have enough space, you can try to steer your horse in ever-decreasing circles. Reach forward with one hand and take a constant pull on one rein only. Using a bit with full cheeks or large rings will aid steering.

It is futile just to haul on both reins unless you have a bit like a gag snaffle which exerts pressure in a severe squeezing action from the horse's poll to the corners of his mouth the more he attempts to take hold.

Steering a bolting horse up a steep hill will tend to knock the wind out of his sails and he can then be driven forward by the rider's legs to continue at canter for a period long after he would choose to slow down if left to his own devices.

If you feel that steering towards a high hedge or fence is the

best option, make sure you arrive at an angle to it and keep the horse's head bent to one side or there may be a risk that he tries to leap it. Do let him see any obstacle beforehand, however, unless you want to crash through it!

Left: If you have space, steer your horse in ever-decreasing circles to channel their speed.

Below right: Some horses will not settle until they have had a 'blast'. Cantering them up a steep hill will help to dissipate energy helping you to regain their attention.

Main points of pressure

This area is shortened as the horse or rider pulls.

Right: A gag bit teaches the horse not to pull. If he takes hold he is effectively pulling against himself and the pressure is released when he stops.

HEAD SHAKING

Problem

We expect to have the occasional problem with a horse resisting taking the rein contact, but a horse that head-shakes will toss his head up violently and may even throw it so far back that he hits the rider in the face. There are other symptoms including sneezing and snorting and trying to rub his muzzle on the ground or a forelimb even while trotting along. Even in its

Above: Riding a head-shaking horse is limiting as you cannot maintain a contact on the reins.
Left: A head shaker is obviously uncomfortable and is trying to alleviate an aggravation. He is not just trying to be naughty!

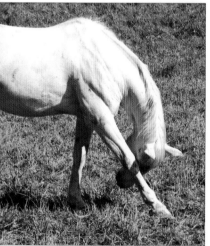

mildest form, just two minutes with your horse snatching the reins out of your hands is an unnerving experience and one that puts a limit on a ridden horse's future. The horse is unable to concentrate on the rider's instructions or his surroundings so it is not safe or pleasurable to ride a the horse in this state.

Head shaking can be the result of a physical problem, behavioural disobedience or may be triggered by a biological discomfort or allergy.

Causes

Unfortunately there are often many factors involved in the development of a head-shaking problem and identifying the trigger can be difficult. It is possible that the horse is simply protesting against a restrictive rein contact but these head-shaking outbursts tend to cease when the pressure on the bit is released. Experience has shown that often the problem worsens in bright sunlight or when pollen levels are high. Horses at rest may be unaffected but when exercise plus the 'trigger' combine, the horse may exhibit incessant or intermittent bouts of head shaking. Another theory is that horses that have previously been exposed to the EHV-1 herpes virus can experience aggravation of the trigeminal nerve. Ear mites, rhinitis (inflammation of the nasal mucous membranes) and guttural pouch mycosis (a fungal infection of the guttural pouch which lies in the horse's throat) can lead to head-shaking symptoms and high blood pressure has been linked to hypersensitivity to skin stimulation caused by damaged blood vessels.

Above: Horses are aggravated by flies and pollen but a prolonged episode of shaking should alert you that he needs veterinary attention.

Remedy

The first thing to check is the fit of the horse's tack, especially the bridle. If the headpiece or browband are too narrow, the restriction this causes (especially around the ears) will aggravate the horse. A horse will often tilt his head sideways or rub his

Equitation and Schooling Problems

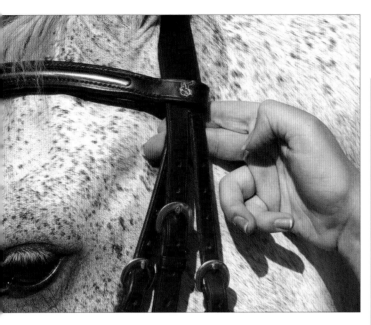

Above: Check the fit of tack especially around the ears. Observe if the head shaking only occurs when the tack is worn and keep a diary of such events.

Above: Gauze nosenets have been very successfully used to limit the amount of pollen entering the nostrils and they can dramatically improve some horses' symptoms.

head which can look alarming but it may simply be that he is itchy from sweating behind the tack at the base of his ears. Check also the width and fit of the bit and make sure that it lies still without cramping the tongue and does not clank against the teeth.

It is important to keep a written record of when the problem worsens.

- Is it only with tack on?
- What is the weather like: overcast and cloud, rainy, bright and sunny, windy?
- What is the date and time of day?
- When does it start?
- How long does it last?
- What makes it subside?

These observations will be invaluable to your vet to create a picture of what may be the trigger, or combination of triggers, if the problem persists.

Many people have found that the problem will be dramatically reduced by using a nosenet – a piece of gauze that affixes to a headcollar or bridle and covers the nostrils and muzzle but which still allows the horse to eat and drink. It serves to reduce the inhalation of pollen through the nostrils and it is possible to actually see the pollen caught in the gauze after one has been worn for a time.

Your vet may prescribe antihistamine treatment to calm the inflammation caused by allergens or a course of B-complex vitamin injections. B vitamins cannot be stored in the body and help to support the nervous system and blood composition.

Inhalation of vapours from the infusion of mint, liquorice, eucalyptus or ginger may help to clear the nasal passages to make breathing easier. Garlic, mint, feverfew, aniseed, chamomile, plantain, marshmallow, nettle and cleavers are all helpful in supporting the respiratory and blood systems to help regulate this debilitating condition.

Adaptogens are natural nutrients that are believed to increase resistance against adverse effects of physical, chemical or biological disturbances. They are derived from plants – *Rhodiola rosea* (golden root), *Eleutherococcus senticosus* (Siberian ginseng), *Withania somnifera* (winter cherry) and *Schizandra chinensis* (Wu-wei-zi – a woody vine from the Far East). They have been used with some success in combating stress and are powerful antioxidants which have been successfully used in the treatment of head shakers. It is advisable to consult a herbalist for the maximum chances of success, rather than attempting to administer without expert guidance any medication yourself to an already sensitive horse.

■ SHYING AND SPOOKING

■ Problem

However well-schooled, many horses have a disconcerting habit of shying and spooking at objects or sounds that they perceive as being alarming, mostly when they are hacked out. A horse can be forgiven if something startling makes him (and his rider) jump, for example, a bird flying out of a nearby bush, but some horses spook at the slightest provocation, sometimes veering violently across the path of another horse or an oncoming

vehicle on the road. The reaction is often far out of proportion to the nature of the 'threat' that caused the upset. It is easy to become unseated or endanger others on such an animal and so this is a far more worrying vice than most of the bad habits that we commonly ask about when viewing a horse to purchase.

Left: Leaping over an unusual obstacle is safe in this controlled situation and this type of training will pay dividends should you encounter 'spooky' objects when out hacking or in the show ring.

■ Cause

A horse's natural instinct is to run away as quickly as possible from anything he fears. Breeding plays a large part too. Highly strung animals are far more on their toes and sensitive in these situations. This isn't to say a heavy cob or warmblood will not shy. Often these types are not actually scared but they use the disturbance as an excuse to misbehave. Some thoroughbred types seem to run on adrenalin and have difficulty in concentrating on the rider's instructions and they come back from the ride dripping with sweat and unsettled. Too much energy from feed, lack of confidence emanating from the rider

and inadequate exposure to 'scary' objects in early training also all play a part in causing an animal to shy.

■ Remedy

A confident rider will exude security to the horse. No matter how much technical ability a rider has, if their confidence has been knocked by a previous spooking experience any tenseness will radiate down to the horse. By employing some of the methods the police use in training their horses we can, with a little intensive training, improve the safety of our horses for years to come.

Use a safe area like an arena or small corral and set up four or five scary objects and obstacles in various locations. These could include a flapping plastic bag, a whirly windmill (such as a toy or bird scarer) and something reflective (use silver foil

Below: The bane of a horse rider's life! Getting a horse past a plastic bag flapping in the hedge can be like trying to ask him to walk through fire.

Below: In an arena you can spend as long as it takes to convince your horse to overcome his fears of unfamiliar objects.

Both above: An assistant can help by acclimatizing your horse to strange sights and also to sounds like clapping or a whistle blowing.

wrapped round a board rather than a mirror which could get broken), all attached to the fence or jump stands. Have an assistant on hand to lead or encourage the horse with titbits if he resists while the rider asks the horse to walk past each object. On no account should the rider or handler get cross with the horse. Talk to the horse and be prepared to stand and allow the horse to take a good look at the object before you ask him to approach it. With moveable objects like plastic bags, your handler can hold the item close to the horse (or even stroke him with it) while feeding titbits and praising the horse. Other useful lessons can include your assistant setting off an alarm or car horn and approaching you with an umbrella raised.

Although contrived, these lessons (and there should be several) will enlarge your horse's field of experience while you are in a safe environment. This should result in the reaction and recovery time being reduced when any unwelcome items are encountered in the future which are out of your control.

It is rare to come across a totally bombproof horse but a horse that settles down rapidly after a fright has a greater sense of self-preservation (and is therefore safer) than one that is totally freaked out and uncontrollable when surprised by something out of the ordinary.

■ STUBBORN OR NAPPING BEHAVIOUR

■ Problem

It is exasperating if your horse suddenly freezes to the spot and refuses to move out on a hack or during a schooling session. Sometimes such horses will literally stand all day rather than move forwards and no amount of cajoling seems to have any effect. This can deteriorate into a situation where the horse refuses even to exit the gates of the stable yard. Often a battle between horse and rider ensues in an effort to make the horse take just one step forwards, and this can escalate into the horse rearing or spinning round and trying to bolt home.

Below: Occasionally no amount of cajoling will encourage a horse to move. Napping can be caused by a lack of confidence or sometimes laziness.

■ Causes

- A horse can be having an off-day and may have either a physical problem or cannot cope mentally with what you are asking at that particular time.
- Horses are naturally herd-bound and take confidence from the familiarity of their surroundings. Some sensitive horses feel insecure away from the confines of their stable or field.
- Other horses have no fear but can simply be lazy or want to head back for their next feed.

■ Remedy

Schooling Firstly, we cannot expect our horses to be keen all of the time. The schooling work may be too challenging or perhaps the horse is tired and achy. If you believe this could be the case, it may be wise to listen to him and face the lesson another day.

Shorter schooling sessions will keep him focused and keen. A bored horse may benefit from a session in the school with another horse. Rather than plodding round in single file, invent a routine working in pairs and in tandem – schooling will be altogether more inspiring. This could be ridden to music, which will lift the mood and assist with rhythm. Using poles and cavaletti (a series of parallel poles set low to the ground on supports) are likely to increase impulsion too.

Riding out Horses are naturally cautious and there may be a genuine reason why he will not pass a certain point. Maybe the ground is soft and he fears that it may give way or he can smell something that he perceives as dangerous. His natural instinct will be to freeze. Give your horse some credit that his motive may be to protect himself (and therefore you). Go out with another horse if you want to explore somewhere new or lead him from another horse to familiarize him with the route. Your horse can then draw confidence from the other animal and may be less reluctant the second time round.

Above: Low jumps are not challenging but help to create impulsion. Riding in pairs side by side in the arena or to music also helps to step the horse up a gear.

Although you may be enjoying the scenery on the hack, he could have had enough and is just trying to tell you that. Physical problems aside, it is not a good idea to let the horse win and dictate when you should go home each time. But subtle tactics that enable you to get the best of the 'dispute' are often better than resorting to force.

Case History My friend Viv would ride our cob Harry and found that he could sometimes be a complete pain when ridden alone. Being an intelligent cob, he was familiar with the all the various tracks that we used to ride and he would stop where the path split and would only move in the direction that was the shortest way home. After listening to him and giving him the benefit of the doubt a few times, she realized that he was just being lazy and she should not put up with him limiting the distance of rides.

Viv discovered several methods by which she could get him to move:

1 Rather than just ambling along she would get him going in an active trot quite a way back from the point where the paths crossed. Shortened reins but, more particularly, the momentum of an active trot helped her direct him from ducking out or planting his feet.

2 Leading him on foot *(below right)* in the direction she wanted worked provided he got rewarded when she was back on board and he was walking on. Do not reward from the ground otherwise you will get back on board only for him to plant his feet again! She rewarded him by finding nice areas of grass along this path where he was allowed to stop for a mouthful before continuing. He actually would then choose this route in preference to the other in order to get the reward. (Treats can be carried in your pocket to reward good behaviour).

TIP If he freezes, try turning the horse's head and steering him in an exaggerated zig-zag. This generally forces him to move to rebalance and this may just break the problem.

Left: If your horse lacks confidence, follow another horse. This applies to the show ring too – enter the ring close behind another horse.

LAZY HORSE – DEAD TO THE LEG

Problem

The horsepower in your horse seems to have vanished. You feel exhausted trying to keep him moving and impulsion is just not a word he understands. This also affects the performance of lateral movements (on and off the horse) if he is unresponsive to moving away from pressure. These horses are not going to be a success in any show ring and are draining for any rider.

Causes

• Breeding plays a large part. A heavy cold-blooded horse like a Clydesdale *(left)* is not going to be as athletic and on his toes as a Thoroughbred or Arab, whose paces are more elevated in any case.
• Just as in humans, lack of fitness and carrying too much weight results in lethargy.
• Horses suffer from lack of motivation too. He may be bored doing the same type of work each day and he may sense your lack of enthusiasm!
• Constant nagging from the rider's legs is more often than not the root cause, meaning that the subtlety of aids has been lost.

Remedy

Temperament: Firstly assess your horse. Take into account his build and overall temperament. If he has sparkle in the stable, he can have sparkle being ridden.
Diet: Is he overweight? Although

Above: Apathy can spread from rider to horse so keep your work varied and give precise aids.

Using the seat to produce a collecting effect.

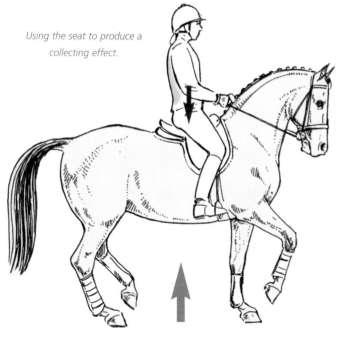

Using the seat to produce a forward driving effect.

Below: Lots of transitions will help increase muscle tone and lightness of bearing.

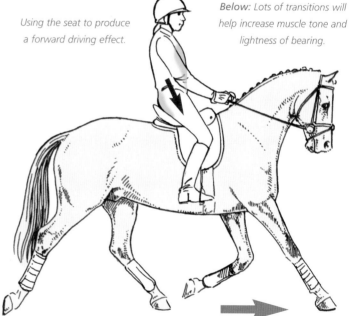

being undernourished will cause him to lack energy, in the majority of cases listless horses are overweight, getting too many calories (known as Digestible Energy [DE] listed as megajoules [MJ] on a feed sack) for the energy they expend. Most leisure horses can work off forage alone so check that you are not being overly generous with his rations – feed for the work actually done, not for what you would like him to do!

Have fun: If your 'get up and go' has gone, you cannot expect your horse to be enthusiastic. Go for fun rides with your friends

(above), play gymkhana games (whatever your age!). Horses pick up the adrenalin aroused by a competitive race and this can be done in the arena or out on a hack.

Be assertive: Give clear, concise aids and don't amble along but think ahead – for example, 'I will trot here' (e.g. at marker A) or 'I will walk for 10 strides and then halt'. Lots of transitions and changes of pace and direction bring the quarters underneath the horse which will aid collection and increase the horse's muscle tone. A half hearted approach – 'I hope we canter somewhere up that end of the school' – will only lead to apathy from the horse.

In the arena or out on a ride it is totally futile to keep nagging with your legs if the horse is lazy or plants himself. You will end up with a horse with bald patches on his sides who becomes more and more insensitive to the leg. A horse's

reward when working is that our hands and legs cease squeezing and this is what maintains impulsion, not continual kicking.

This is the incremental way to make a genuinely lazy horse more responsive.

1 ASK NICELY – squeeze with the legs.
2 TELL – shout 'Walk on' and give one huge 'Pony Club' kick!
3 DEMAND – one short sharp whack with a schooling whip behind your leg aid (remember to change the hand you hold your whip in so that this does not always happen on the same side of the horse).

It is important to be consistent and do this each and every time. The horse will soon learn to respond to the earliest cue that you give. This is far kinder than inflicting ineffective jabs in the horse's ribs at every stride and produces amazing results.

Once you have a responsive horse he will work more effectively, become fitter and therefore more energetic and keen. Exercise produces energy.

On the ground: Establish an assertive role when on the ground. Do not accept the horse leaning on you when you want him to move over or dragging behind when being led. **Ask** with a tug on the reins or a push with the flat of your hand to move him over. **Tell** by prodding him with one finger behind the girth area and shouting 'Walk on' or 'Over' as you do it. **Demand** by facing the direction of travel and using the schooling whip with your outside hand behind the girth area and reiterate the vocal demand.

Above: Be assertive on the ground – remember the sequence ASK, TELL, DEMAND. With practice, you will only have to ASK before you are obeyed instantaneously.

■ JOGGING/TOO LIVELY

■ Problem

A jogging horse can be tiring to ride. A jog is slower than a trot and has much shorter strides. It can look showy and be comfortable for short periods and is a required gait in Western riding, but it is not desirable to have a horse that constantly falls into a jog and will not relax into a walk when asked. Trying to slow the horse down often results in head tossing and just a couple of strides of walk before he resumes jogging.

Above: A forward-going horse can be an exhausting ride if he insists on jogging along like this rather than relaxing into walk.

■ Cause

Assuming that the horse's back, teeth and tack are not at the root of the problem, resulting in jogging as a means of evading pain, then it is most likely that you have a fit or high spirited animal who runs on adrenalin. This is not ideal as he is paying little attention to you, the rider, and will probably finish the ride in a sweaty and tense condition.

When riding in a pair or group, another reason can be because the animal has shorter strides than its companions (e.g. a pony ridden next to a horse) and so compensates by increasing speed rather than extending his stride at walk to cover the same ground.

■ Remedy

Begin by riding him in an enclosed area at all paces using only a light contact. To walk freely a horse needs more rein than to jog. By hanging on to his mouth you heighten the excitement and his outline will be shortened and the impulsion increased. I know my calm plodding cob turns into a coiled spring just by riding him on a short rein.

On a hack **relax**. This is the key. Think calm and don't grip with your lower leg or hands. This does not mean go floppy, however, because if you 'go with the movement' it will make it all the more comfortable for the horse to jog. Hold him with your thighs, seat and back, steadying him with defined half halts, and changes of pace. Make him trot for say 20 paces – and count them out – or insist he does lateral movements (e.g. leg yield to the left, then to the right), then drop back to walk on a long rein. Unfortunately, with this sort of problem repeated attempts to slow the horse down with the reins can merge into constant nagging and may result in a dead mouth and the horse ignoring you. I am a fan of voice commands as they can help define the start of transitions, so include recognizable commands to back up your aids.

Try turning a jogging horse with one rein in a circle while saying 'whoa' and keep on turning until you achieve a walk. Keep your lower legs off him, just turning his head. This helps

Below: Lungeing before riding can take the edge off any pent-up energy and allow him to loosen up and stretch his muscles before you get on.

Far left: Think 'calm' and steady him with your thighs, seat and back. Use half-halts rather than constant pulling and give with the reins to reward each time he slows.

Keep your legs off to disengage the 'engine'.

Left: Steer him in a circle with your rein only and keep turning till he slows before going straight again.

to disengage the hindquarters and therefore the forward impulsion. When you have achieved walk, stroke the horse's neck and allow him to walk in a straight line again. He will probably soon get fed up of jogging if each time it results in him having to turn circles and not being allowed to speed back home. Hopefully this will result in the pressure on one rein (without using the leg) being the cue for the horse to resume the walk. Consistency and repetition will be needed as it is a difficult problem to cure.

A more positive approach is to teach the horse that waiting, rather than rushing, can reap rewards. If you see a nice patch of grass out on a ride, stop and allow a few minutes grazing. This does not mean allowing your horse to snatch at vegetation whenever he chooses, it should be your decision to stop. Do not always stop, walk, trot or canter in the same places and vary the time you take him out and the route and direction of rides so that the horse does not anticipate getting past the half-way point and then charging home for his tea!

Management: More general points to consider include an assessment of the horse's feeding regime and his lifestyle – would he be better on a lower energy, higher fibre diet with perhaps the addition of oil or sugar beet to maintain weight? Often these horses lose condition from being constantly on the go.

If he does not get much turnout, he lacks the freedom to have a gallop at will. How many times have you seen a horse gallop round a paddock, snorting and spinning with delight *(right)*? Have some consideration for his behaviour if your horse

lacks this most basic of needs. With the problem of jogging though, it does **not** help to give the horse a gallop on a hack, thinking the release will help him to settle down afterwards. On the contrary they seem to get more excited and wound up, so it is best to stay in controlled, defined paces.

Tip A fit athlete is on his toes. Take into consideration the fact that your animal may need to expend pent-up energy before he is able to concentrate or relax. Lungeing before riding may be the answer, allowing him to loosen up (and even buck) in a controlled environment, but make sure he has time to wind down before you saddle up and go out.

■ WILL NOT HACK OUT ALONE

■ Problem

Sadly there are many horses with exceptional ability who just seem to fall apart if asked to do the simple task of hacking out alone. They often freeze at the gate and refuse to leave the yard or walk 100 yards and then spin round and charge home. If they do make it out, they will call incessantly and be fractious and jittery which does not make for an enjoyable ride. Owning a horse like this is problematic in terms of the limitations it imposes on your riding and the necessity of having to time outings to coincide with someone else to accompany you.

■ Cause

Horses naturally feel secure in the company of other horses and at their most vulnerable when alone. Although they may enjoy work, leaving their companions and familiar surroundings to venture into the outside world calls for a confident horse. If a horse has not conquered this fear early in life, he can become permanently introverted, only wanting to work in familiar or confined areas.

Right: Riding in company always gives a horse more confidence.
Both below: My mare hates leaving the security of her fieldmates, yet when we ride out with them she strides ahead boldly in front of them.

Equitation and Schooling Problems

■ Remedy

You need to treat these horses as though you were training youngsters. This is the method I would suggest:

- Lead them from another horse or, alternatively, ride out with another horse. This serves the purpose of familiarizing the horse with the scenery and obstacles on your usual trails so these become ingrained and familiar.
- Ride or lead your horse from the ground with an assistant walking or riding a bicycle *(right)* alongside you. The chatting between you will help keep the horse calm.
- Long-rein your horse with your assistant walking or riding a bicycle level with the horse's head, ready to encourage him forward if he has a panic attack. If he is confident enough, ask your assistant to hang back a bit so the horse is taking the lead. Long-reining forces the horse to make decisions on his own – he does not have the encouragement or reassurance of being held between your legs.
- Ride your horse out using the same voice aids as you do while long-reining and with an assistant to help get the horse past 'sticking points' on the ride, but otherwise to hang back.
- When finally you go solo, it can be helpful to arrange to meet a horse part-way through the ride, as horses soon learn that they might meet a friend while out.
- Reward your horse half-way through the ride by allowing him to graze the verge for a few minutes.
- Always choose circular rides otherwise your horse will have a tendency to try to nap home or will refuse to go further than a point at which you have previously turned round.

A good trick is to transport the horse to another location and attempt to ride out from there as there will be no habitual sticking points already established in the horse's mind. Only do this with a horse that is known to load well into the trailer or lorry and carry a mobile phone or you may find yourself stuck in the middle of nowhere! Every positive new experience will expand your horse's horizons and build up his confidence.

Tip You need to have a good relationship with your horse in order for him to accept your guidance. Practising your schooling exercises while out hacking is a useful way of focusing his attention on you rather than the scary surroundings that might otherwise spook him.

Above: *Time spent alone with your horse is invaluable for creating a bond so that the horse will trust you when you ask him to go past a 'scary' obstacle.*

■ SNATCHING AT GRASS OR BRANCHES

■ Problem

It is not very satisfactory if one minute you are riding down the lane nicely, then suddenly you are hauled over to the verge, almost lying on your horse's neck as he chomps his way through the vegetation. To add insult to injury, when you finally drag him away he has half a tree trailing between his forelegs (his takeaway!) plus green slime dripping from his mouth.

■ Cause

Intelligent horses are opportunists – and who can blame them! If I was kept in a stable or a bare field with just hay and my usual bucket of feed, I think I would be attracted by the diversity that the hedgerow offers. It is how horses live naturally after all and horses frequently self-dose with any fresh herbs available to them.

Below: Allowing your horse to snatch at plants on a ride creates an irritating habit as he tugs the reins out of your hands when the mood takes him.

Above: If your field provides trees and shrubs to snack on at home, then your horse does not have the excuse of being deprived during an hour's trail ride!

Equitation and Schooling Problems

■ Remedy

There does, however, have to be a balance between allowing your horse or pony the delicacies encountered out on a ride and good manners. Don't just let him amble along but keep a contact on the reins and keep him between leg and hand. Keep your whip in the outside hand (i.e. nearest the grass or trees) and wave it beside his head or smack him on the neck if he attempts to veer to the verge or snatch at the branches. With children it is helpful to use grass reins on a pony. These are special reins designed to be attached to the saddle and prevent the pony from putting his head to the floor so the poor child will not be hauled over his neck.

Be consistent and insist your horse moves on if he attempts to stop and graze. As a reward for good behaviour, later on during the ride choose a place where there is good grass, make your horse stand first then allow him to graze for a few minutes. You can also tear off tasty branches of hazel or willow and stuff them in pockets or under the pommel of your saddle to feed to the horse when you return home.

Caution Some road verges and field edges may have been recently sprayed with weed killer or pesticides or be polluted by traffic fumes. Only select areas for grazing that are unlikely to be contaminated.

Case Histories I taught my favourite mare to beg for treats by holding up a foreleg. This was very endearing to watch and was a lovely way of communication between us. If I had nothing left to give her, I would tap her twice on her lips and she knew the

Both above: Demand concentration on the ride and you choose a point where he can graze and relax for a few minutes. Make the most of nature's larder and take some vegetation home for a treat later when he is stabled.

treats were finished. If I had forgotten to reward her, she would stop and lift her leg, even with me on board. It reminded me to appreciate what a lovely horse I had. It was not a problem if I had no treats, I could still ask her to continue and the only embarrassing time was when I was standing in line in a showing class and she begged to be allowed her head to graze in the showground!

My clever cob Harry had the memory of an elephant. A year might pass but he would still remember where there was an apple tree on a hack. What a clever horse – unfortunately I could not explain to him that apples were only available during certain seasons!

■ BAD IN TRAFFIC

■ Problem

Sadly today there are few of us who are lucky enough to have our own estate or acreage to avoid the necessity of ever again riding along or across public roads. In any event, it is likely that sometime in his life a horse will be placed with an owner who needs a traffic-proof mount. Car drivers are often ignorant of how reactive horses can be if scared while on the road – if they actually knew how bad horses can be, they would all slow down!

Left: By always thanking considerate drivers and using clear hand signals, let us hope that we can improve car drivers' normal perception of horse riders!

at the last minute causing them to skid and screech in an effort to avoid us – scary!). Another good tactic I have discovered is to ride through a farmyard. Here you may find the biggest and scariest of tractors with rattling trailers while in a safe off-road environment

■ Cause

As with all fears, it is often ignorance and insecurity that have caused initial mild anxiety to escalate to an irrational level. If the horse had been acclimatized to the sights, smells and sounds of traffic early on in his education, he would perceive vehicles as objects to approach with caution but not with blind panic. Of course, a horse or pony that has actually experienced an accident or a 'close shave' in traffic will understandably have a more deeply imprinted fear that may never be overcome.

■ Remedy

While we cannot influence how car drivers behave (other than by being courteous and thanking those that do give us consideration), we can try to mitigate the problematic behaviour of our mounts.

Youngsters or horses known to be traffic shy should be grazed in a field adjacent to a road where they can absorb the sights, sounds and smells of traffic and file them in their brain as a familiar experience. Then progress to riding them along this fence line so that you are confident they will not shy away when ridden.

Alternatively, bring the traffic to the horse. This can be done by playing the horse a recording of traffic noise – horns blowing, air brakes, screeching tyres etc. (we all know only too well how many times people fly round a bend and see the horse

Above: If your horse can be put out to graze in a field adjacent to a road or driveway, this will help to desensitize the fear response triggered by the associated sights or sounds of traffic without putting any rider in danger.

Above: Acclimatize your horse to tractors or lorries in a safe and controlled environment rather than meeting them for the first time on a road.

where you can spend time acclimatizing the horse to the experience of vehicles.

- Educate your horse to traffic in a safe environment before going on any roads.
- The rider or handler should remain calm as any nervousness will be picked up by the horse.
- Go out with a reliable, calm horse as an escort who will give confidence to your mount.
- Do not take risks. If a particularly 'scary' vehicle is approaching do use hand signals and ask the driver to slow down or stop.
- Make use of laybys and lanes to allow your horse to stand at a distance from potential hazards but still keep them in view.
- Keep your horse's head straight or turned slightly to the inside. This will not only help him to see what may be approaching from behind but more importantly prevents him from swinging his quarters out into the path of any vehicle.

Herbal Healing

Before venturing out on the ride, give yourself and your horse a dose of the Bach Flower Rescue Remedy. This should help to take the edge off any fears that may be building up and improve

recovery time after any stressful events you may encounter.

Caution If you ride on the road it is extremely important to have, at the very least, Third Party Insurance. You are putting yourself in a vulnerable position from persons seeking compensation for any injury or damage to property that your horse may have caused and these claims can run into the millions.

Safety Precautions

Wear a fluorescent tabard which states YOUNG (or NERVOUS) HORSE, PLEASE PASS SLOWLY as these have been found to have more effect on drivers than tabards with no message. *(see also Shying/Spooking – pages 118-119).*

Above: A bombproof escort will engender confidence in a fearful horse. I lead all my youngsters from a schoolmaster horse before riding them on roads. A confident and sensitive rider is then needed plus frequent exposure to whatever is the trigger that upsets them in order to desensitize the fear.

■ CANTER PROBLEMS: HOLLOWING/RUNNING INT(

■ Problem

Rather than making a smooth, instantaneous transition from trot to canter, the horse runs on in trot with his neck poking out. This is the typical 'riding school' problem.

■ Causes

- Lazy or tired horse – it requires less energy for the horse or pony to slob along in a trot than to achieve a bouncy canter.
- Poor schooling – the horse is unbalanced and on his forehand.

- Poor aids – the rider is not channelling the power correctly.
- Pain – the horse is stiff and hollowing is a way to evade the pain.

■ Remedy

If, for example, we take the riding school scenario, the horse or pony is likely to have experienced several lessons in succession in which he has taken inexperienced riders round and round the arena. Not only is he likely to be tired but he will have lost

The impulsion is there and this needs directing into collection not speed.

The rider's position is good but the horse is resisting the rein contact.

The horse is hollowing and the forwards impulsion has been lost.

He has thrown his head high as the rider has closed the reins.

The hind legs are coming nicely underneath.

The leg aid is good but the power needs channelling through to the bit.

Rather than channelling the speed, the rider has panicked and tilted her weight back, blocking the horse.

The rider's legs are forward and not asking for impulsion or engagement.

Equitation and Schooling Problems

Weight mainly on the forelegs

Centre of gravity moved backwards

Right: Horses naturally carry 60 per cent of their weight on their forelegs and so need to be schooled to move their centre of gravity further back. Far right: The rider has created impulsion stemming from the hind legs and harnessed it with her rein contact. This channelled energy causes the horse to step further underneath him and creates more elevated paces. The shaded areas show how the centre of gravity has moved.

motivation to maintain that 'spark' that facilitates accurate transitions. It takes a good rider or a forgiving horse to generate the impetus.

Horses naturally move 'on the forehand', with most of their weight being on their front legs. Classical riding asks for them to carry more weight behind to achieve elevated paces and lightness in the reins. Horses need to be taught this and schooled in this manner to maintain muscle tone. It does not always come naturally, it is tiring for the horse in the early stages of schooling and its conformation may impede the action (e.g a stocky, long backed Welsh Cob with a naturally choppy action will find it harder than an Anglo Arab with naturally more elastic paces). A poorly fitting saddle will also cause a horse to hollow.

The rider needs to channel the impulsion from behind into the rein contact. It is easy to want to kick with the legs while allowing with the hands to achieve speed but this will not produce the strike off that you need and the power will just go out the 'front door'. Keep your reins short and use a half-halt by checking with the outside rein to get the horse to listen to you then squeeze with your legs and gather up this power in both reins. A collected canter may actually feel harder to sit to as the movement is slower and more 'springy' and harder for the rider to absorb. Achieving the correct lead is discussed overleaf.

Below: The left rein provides the steering function while the right rein checks the horse before asking to canter. The leg then should maintain a contact to stop the horse falling in on the circle. Sit deep in the saddle and squeeze with your left leg on the girth and ask for strike off with your right leg behind the girth.

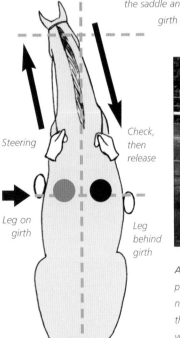

Steering

Check, then release

Leg on girth

Leg behind girth

Above: Breeding and type plays a part. An Arab or Lusitano will naturally have more elastic paces than a draft or driving type of horse which will have a choppier stride.

■ HARD TO INSTIGATE THE CORRECT LEG AT CANTER

■ Problem

It is desirable to be able to choose the leading leg in canter according to the direction of travel. Although it is a hind leg which instigates the start of the canter transition, we refer to the 'leading leg' because a rider can glance down at the forelegs to check they are on the correct lead. If you are going to the right on a circle (i.e the right rein) the inside ('off') foreleg should appear to sweep further forward than the outer ('near') at canter. If you are on the wrong leg it will feel and look unbalanced and the horse will find it much harder to make a smooth turn. When showjumping, being able to land after a jump straight into canter on the correct lead for the direction of the next jump will be advantageous in terms of speed and agility. The problem comes when a horse constantly favours one particular leg or chooses the wrong leg which would lose you marks when performing an individual show or dressage test. There is a movement called 'counter canter' in dressage but, again, the rider needs to be able to demand the desired lead at will. Another common problem is when the horse is 'disunited' – although he appears to follow the correct lead, actually his hind legs are opposing this and moving on the opposite lead. This generally happens when the rider has unsuccessfully attempted to correct the problem.

■ Causes

- A horse that is generally more supple on one rein than the other – pain, stiffness, uneven muscle tone can all cause this.
- If a horse is rarely schooled and predominantly hacked out, it is likely that the horse, rather than the rider, will determine which leg to canter on and may habitually choose the same one.
- The rider may not set the horse up correctly for the correct bend – muddled aids.

■ Remedy

It is hard for a young or unfit horse to canter on a tight circle with a rider on board as he needs to place his inside hind leg well forward beneath him to balance himself. Suppleness and strength can be improved by lungeing without a rider. Always

Above: This picture shows an excellent horse/rider partnership. The horse's hind legs are working well and the left leading leg is obvious.

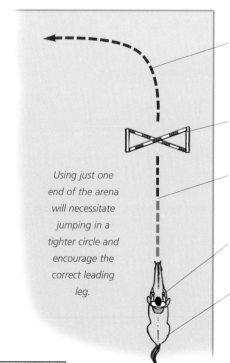

Sit up on landing and encourage the horse to continue in canter.

Lean forward and give with the reins over the jump.

Using just one end of the arena will necessitate jumping in a tighter circle and encourage the correct leading leg.

Squeeze left leg on the girth, right leg behind girth.

Sit to the trot and collect with your reins.

Ask for an active rhythmic trot.

Above left: The horse has landed after a clear moment of suspension. However, the weight should be more on the hind legs.

Above right: This photograph captures the canter sequence of the leading front left leg, then the diagonal pair, and finally the outside hind leg.

start on your horse's better side and warm up beforehand with lots of walking and trotting on both reins.

Place a pole on the circle raised about 30cm (1ft) off the ground. The horse can trot up to it but ask him to 'Canter' as he is about to take off and insist he canters on landing. Many horses will naturally land in canter on the correct inside leg and certainly once a horse has done it correctly, he will feel how much easier it is to circle in canter on the correct lead. As with all exercises, make sure that you devote an equal amount of time in both directions. Regular work like this will even up muscle wastage caused by one-sidedness or a rider not sitting centrally.

When you subsequently attempt to canter him with a rider on board he will be clearer about what is expected. Start with larger circles until he is feeling balanced and always ask for canter as you go into a bend or, as you did with lungeing, as you land after a small jump.

The aids need to be clear and concise or they will muddle the horse (this is where we find the problem of a horse being

disunited in canter). You need to start with an active rhythmic trot. Just before you ask for canter, sit to the trot and slightly check the horse with the outside rein (the pole serves the same purpose of being a bullet point to check the horse and make him concentrate). Sit up, keep your eyes up but looking slightly in the direction of travel, bend his neck slightly to the inside, squeeze with your inside leg on the girth position, and move your outside leg back to stop his quarters swinging out. Holding the impulsion between leg and hand is the key – reins like saggy washing lines will cause the horse to run on at a faster trot. You are looking for a 'spring' from trot to canter. If he does go onto the wrong leg, slow him down and do not ask again until you are back in a steady rhythmic trot and always as you go into a corner.

If he is fine on the lunge but you still have problems, ask an experienced rider to try. It is worth considering whether the horse may suffer from musculoskeletal pain or perhaps arthritis (in the hock) which may be hindering his strike off or problems with his neck vertebrae that are hindering flexion.

■ WILL NOT MAINTAIN AN OUTLINE OR 'GO ON TH

■ Problem

Certainly in classical riding there is a presumption that a good horse and rider combination is demonstrated by the horse working in the correct outline to achieve lightness. This takes time to achieve as the horse needs to strengthen the muscles along his neck, back and quarters (known as the topline) as well as learning to shift the centre of gravity back in order to lighten the forehand. The problem is that many horses either resist working like this and slob along on the forehand or, in our keenness to achieve this at all costs, are merely restrained into a position of having their head tucked in and thereafter simply fix themselves in this position rather than working properly from behind.

- Rider is fixing the horse's head and trying to ride from hand to leg, rather than from leg into the rein contact.
- Wrong muscles have been built up through incorrect schooling.
- Over-bent horse evading rein contact.

■ Remedy

Rather than simply thinking of the picture or outline you want to achieve visually, you need to understand how the horse should be working to achieve self-carriage and lightness when, naturally, he would be predominantly working with his weight on the forehand. Although some horses appear to go into an

Above: It is almost as if the front and back ends of this horse are not connected at all and the horse is just 'slobbing' along in trot.

■ Cause

- Unfit or immature horse.
- Elderly or stiff animal or one with a physical problem.
- Poor conformation.
- Not enough impulsion.

Above: The horse is hollow and not tracking up into the footfalls left by his front hooves. There is no communication from the rider.

'outline' when they are loose, this is usually display behaviour when they are excited and more common in hot-blooded horses. Do bear in mind that poor conformation (e.g. a long back and neck with short legs) can drastically affect the horse's ability to achieve a good outline, as will different breed profiles.

Above: A horse that is tracking up nicely will place his hind legs in the hoofprints left by the front hooves.

Above: Allow your horse to stretch fully to 'warm up' and 'warm down' before and after asking for prolonged periods of collection to release tension.

Horses with thick jowls will find neck flexion difficult and a horse bred for driving with a choppy gait may find it hard to reach out.

If you ride an unmuscled or unschooled horse, he will naturally have a flat or hollow topline and will stretch out his neck to achieve balance. He is likely to lean on the reins and take short, choppy strides. Even in Western riding where the outline is very different to that expected of a classical style, it is still desirable to achieve some degree of collection and lightness and it is much harder to disguise a false outline if riding with a shank bit and a long contact.

Right: Draw reins have their place in training if used sympathetically to aid vertical flexion. However, it is essential that the horse works from behind and does not simply have his nose hauled into his chest which achieves nothing.

A good outline is derived from the impulsion that is created by the rider encouraging the horse forward with leg aids, but also harnessing this power rather than allowing the horse to release it forwards into a faster gait. Many horses will not go on the bit properly because they are not actually going forwards actively – you cannot harness the power if there is no power there in the first place! Therefore a tired or lazy animal will not step out well. This is easily identified by seeing whether the hooves of his hind legs land on the same marks as his forelegs or whether they are 5cm (2in) or so behind them. This is known as tracking up. If he is not tracking up into those footfalls, he is not using his hind legs properly. This could be due to lack of muscle tone, stiffness, pain from the saddle or some physical problem that is far more serious, so the cause should be investigated.

continued ➡

■ WILL NOT MAINTAIN AN OUTLINE OR 'GO ON THE

Firstly, work the horse on a long rein to allow him to limber up and stretch out. If you immediately take up a contact on the reins, a horse is likely to become tense and you want him to feel loose and supple. Get him going forwards well. Have a very loose contact and if he gets too keen and speedy, use circles to settle and steady him but, again, using your weight and seat for steering, rather than by hauling him round by one rein. Get the horse listening to you with lots of changes of direction plus transitions from walk to trot and back again. It is important that he is allowed five to ten minutes of warming-up time and is allowed to stretch out before asking for any collection.

Below: Work long and low to begin with to loosen up the horse and allow him to extend his frame into the rein contact. Get him going forwards actively and try to feel his back swinging as he moves.

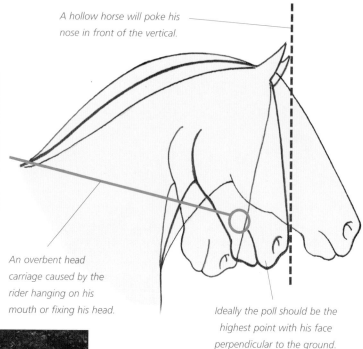

A hollow horse will poke his nose in front of the vertical.

An overbent head carriage caused by the rider hanging on his mouth or fixing his head.

Ideally the poll should be the highest point with his face perpendicular to the ground.

Above: It is important not to dwell on the position of the head as the main criterion of achieving collection. The head position should derive from the horse taking the power forwards from impulsion achieved behind into a light contact with the rider via the bit.

Once he feels relaxed and is moving with a low swinging action, you can gradually take up a contact but keep it elastic or the horse will fix against you or lean on the bit. You want to continue to ask him to move forwards actively but now he can only reach down and forwards as far as your hands allow, so the energy is condensed into a smaller frame – a bit like a concertina. Turns, circles and half halts will help this. You should feel his back lifting under you and his neck flexing. His hind legs will step further underneath him and the stride should feel more springy and elevated,

Above: Lots of changes in direction and transitions from walk to trot and back again will encourage the horse to step underneath further and take more weight on the hind legs which will produce lightness and elevation.

Above: This expressive trot shows how there is fluidity in the rein contact and the horse is relaxed with pricked ears and loose back and tail. There is no sign of resistance or hollowing (compare with the pictures on page 136).

almost as if you are moving in slow motion – a fantastic feeling when you achieve it!

The visual outline with the horse's head to the vertical comes from the arc of the horse – from his hind legs, through his back and to the contact point at the bit. The horse is working from back to front so you can appreciate that trying to force his head into the desired position will actually hinder him working correctly. A well-schooled horse will still hold his outline when the rider releases the reins as he is not leaning on the bit but should be balanced through the core of his frame.

A horse that is 'over-bent' will have the front of his face behind the vertical and the arch of his neck will be the highest point. This is normally caused by the rider having too fixed a

rein contact. The horse is evading the bit rather than seeking the contact. Ideally the front of the horse's face should be vertical or just in front of vertical and the highest point should be at the poll. Taking and allowing with one rein at a time encourages the horse to seek the rein contact. If you get someone to video you, then you can compare it with the outline achieved by a dressage rider and see how their horse is moving and if the picture is one of acceptance or resistance.

Remember that it is extremely tiring for a horse who is not used to using these muscles to work in this way so gradually build up the duration of each session. If your horse is lazy, try schooling him while you hack out if he is likely to be more forward-going on these occasions than when in the arena.

■ DISTRACTED AT DRESSAGE

■ Problem

Schooling forms a large part of life with a horse for many people. Whatever your goal, from showing to eventing, all horses will benefit from time spent schooling in the arena to improve muscle tone and self-carriage and to provide the groundwork for all disciplines. It can be frustrating if you are prepared to invest this time and energy but your horse just does not want to concentrate. Many horses will toss their heads, fall out of a good outline and get increasingly grouchy the more you do. You may feel that they will never be able to perform a good dressage test. Unfortunately riding demands teamwork and compliance between horse and rider.

■ Causes

- The horse is stiff, unfit or unsound.
- Soured by too much of this type of work.
- Affected by too much mental pressure.

■ Remedy

Horses generally want to please. If you have a good relationship with your horse but he resists any sort of training, then perhaps there is an underlying problem. He may start the session well but pain or stiffness could increase the more you ask of him. On a hack these problems are generally less obvious as you are moving forward in straight lines (and usually not attempting to maintain an outline for a long period), therefore the demands on the horse being asked to circle etc. in an arena will be far greater. Ask a horse chiropractor to check him over.

Many horses spend far too long in stables. It cannot be much fun for them to come out of the confines of a stable just

Below left and right: It is frustrating if you are prepared to invest time and energy into perfecting a movement but your horse has different ideas. A perceptive rider will be able to identify the cause of obstinacy and know whether to ignore it or to listen to the horse.

Above: Asking for self-carriage and engagement or tight circles may show up signs of pain or stiffness which are not apparent on a hack. Mentally it can be hard for some horses to understand what is being asked by the aids.

break, plus a stroke or titbit) at the smallest sign of progress. Always end on a high note, so if things are going badly, go back to doing something you know they can manage.

Tip Remember to **warm up and cool down** the horse every time to prevent strains and stiffness.

Above: Turning the horse out to graze and enjoy time just 'being a horse' will help him to relax and be able to cope with periods when he has to concentrate.

to be ridden in another confined space – the arena. With so many sights, smells and sounds coming at them from the outside world, it is understandable that they may choose to focus elsewhere rather than on perfecting a figure of eight or a half pass! Cross-training is essential for the well-being of the horse. Expanding his horizons and experiences will help across the board, whatever the career at which you are aiming him. Turn the horse out to graze more and to socialize with other horses. A happy horse who is relaxed and can enjoy some freedom will not mind the confinement and mental pressure if it is just part of his life. You can always do some schooling while you are out hacking, it is just a case of being more inventive with your sessions.

Some horses have naturally poor concentration spans – I have known several Arabs who get very agitated after more than half an hour in the school. Short sessions are the key. Go over things they are good at and reward them (by allowing them a

Right: Keep training sessions short and allow breaks on a loose rein to refresh the horse before moving on to the next movement. A break and a stroke is nearly always the best reward for the horse.

■ MOTORBIKING ROUND CORNERS

■ Problem

It is quite a common occurrence to see people attempting to school in an arena having problems at canter. Rather than there being the lightness that we seek, the horse or pony is careering round each corner at nearly a 45° angle and cantering a full circle resembles riding 'the Wall of Death' on a motorbike!

■ Causes

- The horse is too much on the forehand.
- The horse is stiff in his back and evading having to bend.
- The horse is unbalanced with the rider on board.
- The rider's aids are not asking for the bend correctly.

■ Remedy

Horses have to be taught to carry themselves, plus the additional weight of a rider on board. If you imagine trying to carry someone on your shoulders, it would be relatively easy to run forwards in a straight line but very unbalancing trying to keep your rider perpendicular if you tried to circle quickly round a corner.

Stiffness can manifest if there is muscle wastage or pain. A bulky saddle can restrict the natural motion of the horse's shoulder and compensating for a rider who sits heavier on one seat bone will, over time, cause problems. It can be very helpful to get someone to watch you riding on the lunge. They can pay specific attention to your riding position and any external influences that may be impeding fluidity of the bend.

It is essential to encourage a good bend and suppleness along the horse's entire length. Tugging the horse's head

Above: Raising your inside hand and sitting upright will lift the horse's head and make him less heavy and on the forehand. It also helps to get him bending and looking in the direction in which you are travelling.

Right: A motorbiking horse is unbalanced and evades having to bend which could possibly be a sign of stiffness. It feels very unsafe when turning between jumps.

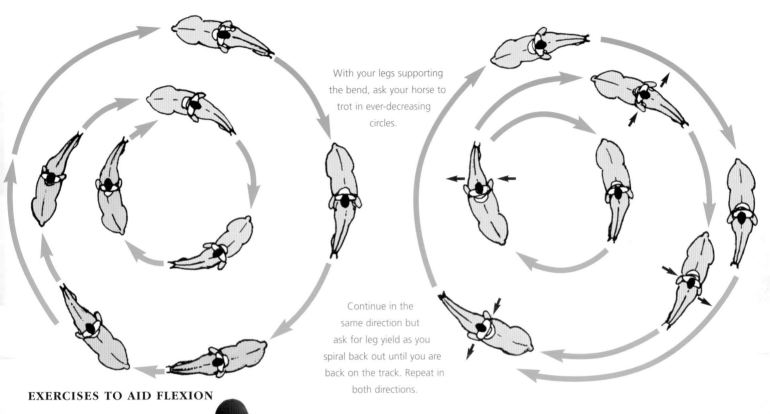

With your legs supporting the bend, ask your horse to trot in ever-decreasing circles.

Continue in the same direction but ask for leg yield as you spiral back out until you are back on the track. Repeat in both directions.

EXERCISES TO AID FLEXION

Above: Once you have achieved balance and bend in trot, practise cantering circles and figures of eight. Remember to sit up and lift the horse with the inside hand and with support from the legs.

Left: You can really see true bend from nose to tail on this pony.

round the corner without adequate support from your leg aids is likely to produce a 'motorbiker'.

- Squeeze and raise the inside rein (by about 15cm/6in) to direct the horse without him falling in.
- Hold the outside rein to check the speed and stop the horse from cutting the corners.
- Keep your inside leg on the girth to encourage impulsion.
- Put your outside leg behind the girth to encourage him to bend his entire body.
- Sit up straight, keep your head up and do not lean inwards.
- Hold him between hand and legs to stop him falling on his forehand.

■ JUMPING PROBLEMS: SNAKING ON APPROACH

■ Problem

When you ride a course of showjumps it is unnerving if your horse will not approach a fence straight but snakes between jumps with his head in the air and his feet shuffling so you cannot get an even stride. It makes it very difficult to predict his take-off point or settle him after each fence to set him up for the next one.

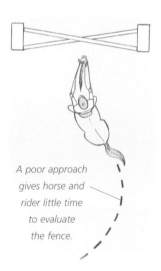

A poor approach gives horse and rider little time to evaluate the fence.

Left: A horse snaking between fences is backing off. This shows a lack of confidence which could stem from simple inexperience, poor coordination or from some physical cause, such as a reluctance to jump through pain or poor eyesight.

■ Causes

• Immature or inexperienced horse.
• Unable to judge distances properly.
• Uncoordinated.

■ Remedy

Although horses have the biomechanics to jump well, they do not usually encounter a series of coloured jumps to leap over in their natural state! Showjumping is a skill that has to be learned and practised in order to develop the confidence and technique to be proficient at it.

A horse that approaches a fence in a snaking motion is actually 'backing off', trying to give himself more time to assess what lies ahead of him. He is not being naughty yet, but will require a confident rider who can hold and guide him towards the jumps to avoid this habit escalating into napping or refusing. It may be helpful to approach the jumps in trot. This will give him that extra time he needs to see the fence properly and get his legs coordinated in time. Remember horses do not see the colours on fences as we do and some colours will

Above: Help your horse gain confidence by setting him up for take-off and keeping him balanced between fences without restricting his rhythm.

appear nearer or further away to them. A horse needs to be facing straight at an object to focus properly on it and will lower or raise his head to sharpen the definition of the image he is seeing.

Regular lungeing or loose schooling over fences will improve his skill and encourage a horse who has previously been overfaced or lost confidence. Keep related distances between jumps simple and tailored to his striding.

LENGTHENING OR SHORTENING STRIDES

Riding over and through grids of poles on the ground will make him more aware of his legs and where they are. This is a very useful exercise for horses with poor coordination and can be done as part of your general schooling.

Right: Schooling your horse over fences on the lunge is a great way to gain his confidence. He can encounter a variety of fences and related distances and the onus is on the horse to set himself up correctly.

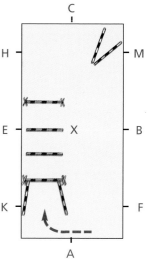

Above and left: Many horses are clumsy and inaccurate. Riding through grids and simple courses of of poles on the ground forces the horse to be more aware of his legs and really helps with coordination.

■ JUMPING PROBLEMS: REFUSING/RUNNING OUT

■ Problem

These are common problems which we will all face at some time but a horse that is consistently bad is not going to endear itself to its owner nor win any prizes in the showjumping ring.

■ Causes

- A previous owner may well have smacked and booted the horse or jabbed him in the mouth and the ingrained memory of jumping is not a good one.
- A tired, stiff or unfit horse will not jump smoothly.
- The horse may be overfaced and has lost confidence.
- General lack of jumping aptitude.

Above: Refusing is a worse habit than knocking poles (where at least the horse has tried to jump the fence) as he is really saying 'No'.

Above: Ouch! Landing in a heap on a pile of poles is going to hurt horse and rider and may well dent the confidence of both. After such an incident it is best to let the bruises heal and go back a level to boost confidence.

■ Remedy

The skill of jumping needs to be taught and encouraged by a competent rider who is sensitive to the horse's faults and fears. Horses can, like people, have off days. They may be feeling under the weather or be stiff from the work done the day before. Give your horse the benefit of the doubt if he plays up rather than resorting to force which will only serve to instil a dislike of being made to jump.

You may be overfacing the horse. Lower the fences and design them to be encouraging with wide wings, and spreads consisting of poles graduated upwards (always with a ground line). Make a concerted effort to stay with the movement of the horse so as not to pull him in the mouth. Do not use martingales when jumping as these can impede the horse's movement and cause him to hollow or become stiff. Another thing to consider is that perhaps you are too heavy for the horse or pony for this sort of exertion.

For horses that run out, keep a good contact but keep your hands wide apart and low to allow the horse to see the fence properly. If his head is pointing in the air he will not be able to focus properly. You will need quick reactions with your leg aids to keep him straight.

To encourage the horse's enthusiasm, ask someone to ride ahead of you over the jumps as your horse will be much more likely to follow someone else's lead. Taking a horse cross-country (ideally as one half of a pair) is great fun for both horses and

Right: This martingale is constricting the horse even when his head is in a normal position! Using a martingale when jumping is likely to impede movement and make your horse hollow.

riders and even those who are dreadful in the showjumping arena can show great ability when they are enjoying themselves out on a cross-country course.

With any jumping problems, consider that the horse just may not be suited to a career that is focused on showjumping. Perhaps his aptitude lies elsewhere, in which case he may not be the horse for you and will enjoy a more successful future career with someone who does not set such store on jumping ability.

Above: Taking a horse to a cross-country schooling day or 'pairs' competition, or perhaps a Team Chase, should renew a stale horse's enthusiasm for jumping.

Standard distances that are used for building a show-jumping course are based on the average length of a horse's stride at canter, 12ft (3.65m). A related distance is the distance between fences. A good rider will walk the course and assess whether they will need to push or steady the horse to get the best take off.

Bounce – 12ft (3.65m)
One stride related – 24-25ft (7.3-7.6m)
Two stride related – 33-36ft (10-11m)
Three stride related – 45-48ft (13.7-14.6m)
Four stride related – 60ft (18.3m)
Five stride related – 72ft (21.9m)

An upright fence will make your horse jump in a tighter outline than a spread fence where he will reach out and cover more ground on landing, so bear this in mind when assessing where you will land on a double or combination fence.

■ JUMPING PROBLEMS: CAT LEAPING

■ Problem

Some horses have the most horrible technique of jumping. They will get close to the fence and then do a huge leap (often far higher than needed) which will often throw the rider forwards or out of the saddle. There is very little flow when tackling a course of jumps as the horse will often break pace to gather himself before the next leap.

■ Causes

- An immature or novice horse.
- Lack of coordination or unbalanced.
- Fear of jumping.

■ Remedy

As stated above, horses need to learn the skill of jumping through practice but with a horse that cat leaps every jump may scare the animal still further if he is not enjoying it. Firstly, check the fit of his tack and only attempt jumping on a horse that is relatively fit and well muscled. Jumping puts an enormous strain on the limbs and requires adequate suppleness to bascule over the fence.

It is helpful to lunge or loose school the horse over low fences to gain his confidence. By putting two or three low parallel poles next to each other, you can encourage him to reach out, rather than upwards. Without a rider he can devote his full attention to weigh up the height of the fence and the best place

Right: Schooling your horse at a faster pace over a series of low or spread fences will encourage him to lengthen his stride and reach out.

Below left to right: This horse has caught his rider out; she has been left behind and has jabbed him in the mouth. The rider must try to instigate the take off and stay with the horse's movement throughout.

Keep looking ahead in the direction of travel.

Keep your upper body parallel to the horse's neck and give with your reins.

Tip forwards – and put your weight into the balls of your feet and your knees.

Judge the take-off position to give your horse the best chance of jumping fluidly.

from which to take off. The more proficient he becomes at this without a rider, the better he will be when under saddle.

When you ride a horse that has a tendency to cat leap, grab a handful of mane or use a neck strap to avoid pulling the horse in the mouth or running the risk of unbalancing him on landing if you get left behind the movement.

Remember: For a good jump, a horse needs to lift his back, tuck up his legs and then stretch with his neck slightly downwards for a fluent landing. If you try to haul him upwards with your hands, he is likely to hollow rather than bascule, which requires him to be loose and supple through the back.

Above: This horse and rider are in perfect harmony. You can appreciate the elegant elongation of the horse's frame from nose to tail and his fluid (rather than stilted) movement.

■ RUSHING FENCES

■ Problem

Horses seem to divide naturally into those that love to jump and those that don't. Horses can certainly be taught to jump but you are never going to have a top showjumper if he is suspicious of every obstacle he approaches and starts to back off. Good jumping horses are valuable and much sought after and many prizes are won by the 'point and shoot' type of horses or ponies that know their job.

Unfortunately you need good balance and nerves of steel to ride this type of animal and when it comes to the crunch, say a tricky jump-off in a competition, will such a bold horse be able to negotiate a tricky short-striding course plus coping with the added height? If your horse is permanently on full throttle and not listening to your commands, you will have a job even to steer the animal at the correct fence. I have seen many frustrated children and adults whose ponies or horses have forfeited a top prize by tanking round the course and missing out one of the fences altogether. Use the warm up arena to get your horse listening to you.

Excessive speed at the point of take-off causes horses to flatten rather than bascule over the fence. You may get away with it while on a low cross-country course (and indeed do extremely well at lower levels), but when it comes to negotiating more demanding fences which require an accurate line/striding or just a good deal more effort in achieving the height, you and your horse may find yourselves in trouble.

Indeed, when trying to retrain this type of horse, practising over showjumps rather than solid cross-country fences will be a far more constructive discipline, as you will know if your horse has flattened out and just skimmed the pole. Fortunately in such a case, it is likely that it will only be the pole that comes down and not the pair of you!

Left: 'Point and shoot' horses are useful but it is a constant battle for a rider to steer a horse on full throttle. His keenness can mean that he sometimes flies past the jump he should have negotiated.

Right: A tricky turn can easily catch out an unbalanced horse. Nerves play a part so try to think 'calm' and ride a route that gives him time to see each fence.

■ Cause

Excitability

Many horses genuinely enjoy jumping and are just enjoying the physical and mental challenge of the sport.

Fear, pain, confusion

A horse or pony may have experienced aggressive riding in the past to get him round a course of jumps or he may find the obstacles themselves overwhelming and just wants the experience to be over as quickly as possible.

■ Cause and Remedy

Cause: Poor management regime. If your horse is stabled for most of the time and only comes out for schooling sessions in the arena, he may see jumping as a time when he can kick up his heels and expend some real energy.

Above: Get rid of some of the pent-up energy and gung ho attitude by hacking out first or jumping on the lunge before asking for concentration.

Right: Tie-downs and martingales cause tension, stiffness and restrict the fluid movement that we seek. A horse needs to be confident that he can stretch out and bascule unhindered – so why are these gadgets commonplace?

Remedy: Schooling beforehand is **not** the answer as the energy is too 'contained' and the explosive power is coiled like a spring. Try lungeing the animal first or ideally hacking out (including an unrestricted canter) and then come in the school afterwards for jump training.

Cause: Fear or pain. Sometimes horses rush fences in order to get round the course as quickly as possible. Riders suffer from nerves and excitement as well and they may boot their horses in the ribs or haul them in the mouth. Once out of the ring, all is back to normal again and the horse can relax.

Remedy: Ride as calmly as possible – don't just flap your legs or use spurs in an attempt to get round the course. You want your horse to trust you and wait for your commands. Developing your relationship on the ground will help here, for

Above: When lungeing over a fence, do not let your horse get overexcited but keep him listening by asking for lots of changes in pace. The jump itself is merely an 'incidental' part of the exercise.

instance, by leading him over a single jump from the ground and then turning his attention back to you.

Cause: Problems with tack. I cringe at the number of times I see flash nosebands (tie-downs) and martingales on jumping horses. Some horses naturally need to open their mouths when exerting energy and tie-downs, in my opinion, hinder rather than help a horse take a good contact in any event. A martingale would also be a hindrance in the showjumping arena – just as the horse takes the leap into the air and tries to stretch over the jump he could receive a yank from the action of the martingale – what an unnecessary encumbrance.

Remedy: Check the general fit of all tack used and **take off anything that may hinder rather than help a free action.** *(See the sections on **Bolting** and **Crossing Jaw** for more advice).*

continued ➡

RUSHING FENCES: JUMPING EXERCISES

You want to encourage the horse not simply to decrease his speed round the course but, more importantly, to increase his focus and concentration on each obstacle and the rider's commands.

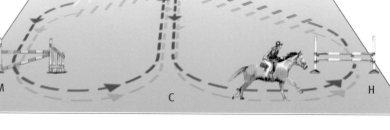

Build upright fences

Upright fences such as walls, gates and oxers with no ground line will encourage the horse to shorten his stride on the approach to the jump and get closer to the fence before leaving the ground at a steeper angle. Spreads have the opposite effect of encouraging the horse to jump on a longer, lower stride so they should be avoided when training horses who rush.

Above: On landing, sit up and leg yield in trot across the diagonal, only allowing a stride before the next fence to go back into canter.

Build courses which require skill not speed

Courses which include several changes of direction will require accuracy on the part of both the horse and rider. Changes in striding lengths (making stride distances shorter than the standard stride in a combination fence for example) plus using bounce fences should make the horse's striding shorter and more rhythmic.

Below left to right: Build courses which require the horse to do a 'double take'. Unusual layouts or alignments plus bold fillers and narrow 'stile' type obstacles will keep your horse focused, and not rushing blindly ahead.

Build fences which require a 'double take'

A course of standard poles and jump stands is likely to be undertaken with more confidence (and therefore speed) than a course containing, for example, brightly patterned planks, a narrow stile and poles set at odd angles to the horizontal. These require the horse to assess each obstacle rather than plough on at full speed.

Combine schooling and jumping

Often a rider's ability seems to go to pieces when jumping and the accuracy and sensitivity used for schooling and dressage is lost. Many people suggest that trotting up to jumps or circling

Above: The leg yield is a good exercise to get your horse listening to you. The energy is contained and you are asking for core strength and balance.

before and after jumps will curb speed and these tactics can be useful. However, I feel that further exercises are required to try to cure horses which anticipate speed and direction of travel by means that will redirect their attention back to the rider and prevent the rider from simply hauling on the reins in a bid to slow the horse.

A good exercise for both horse and rider is 'Jump – Leg Yield – Jump' as illustrated on page 152. Set out four fences close to the corners of the arena and tackle them as a figure of eight pattern. Jump the first fence, leg yield left across the school to just in front of jump 2, then continue around the arena to jump 3, leg yield across to fence 4 and come up the centre line and halt at X. This can be done in trot

Right: Poles used before and within jumping grids will create a 'pause' where the horse has to pick up his feet and adjust striding, meaning that the speed is curtailed.
Far right: Now which jump shall I take? Vary your route to keep the horse thinking.

(or in canter changing the leg across the diagonal) and on both reins and should increase the horse's lightness and concentration. Another version is to position jumps as a serpentine figure.

By altering the sequence and direction in which you take fences it is possible to use one course of jumps to teach your horse that he cannot simply rush on to the next jump in his line of vision but may have to change direction (and therefore canter lead). This means he must listen to your aids first.

If you include trotting poles between the fences and other obstacles such as narrow corridors and fences that can be jumped several different ways, you should be well on your way to making your horse accustomed to problem-solving and listening to your aids for guidance.

Artificial Aids – Bits

Bear in mind that if you are lucky enough to have a horse that enjoys jumping, every attempt should be made to encourage this enthusiasm. Constant pulling on the reins, or using a stronger bit in an attempt to curb this exuberance, may have the effect of putting him off jumping altogether, resulting in refusals or bad behaviour. If he is a good jumper but you cannot stand the pace, then perhaps it is better to let someone more experienced ride him and take pleasure in watching instead.

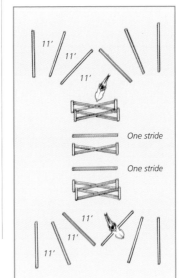

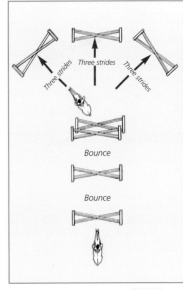

■ THE IMPORTANCE OF CROSS-TRAINING

If you only were allowed to read books on fishing life would be dull! In the same way, just because you are really keen on dressage or jumping, it does not necessarily follow that your horse is! He may soon tire of being dragged out of the stable to go over and over a certain dressage movement or a course of jumps. Keeping his interest is of utmost importance. Cross-train your horse in all disciplines, get out and see new scenery on pleasant hacks and try other things – Le Trec, distance rides, cross country etc. You may also discover his talents lie elsewhere.

Even the most expensive horse with incredible competition ability can have debilitating flaws in other areas. When you try out a horse before purchase it is easy to be impressed by his jumping or dressage aptitude but a high price tag and spectacular specific talent does not always translate into him being an equally straightforward horse on a daily basis.

Above: Try something new like Le Trec which is fun and accessible to riders and horses of all abilities. You will enjoy a great sense of comradeship.

Case History

I once owned a superb horse who was so well-mannered in hand and well-schooled under saddle that a young child could have performed advanced dressage with ease, but if he was hacked out or asked to compete in strange surroundings, he would become unpredictable, uncontrollable and lose any notion of self-preservation. You would not have found this out at any trial if the horse was on his own territory and it is quite possible that the competition-minded previous owners had not had an issue with this if they limited the type of work done with him. So remember, let the buyer beware!

It could have been that his early training was geared to schooling and involved very little cross-training. His ability was recognized and his future duly mapped out. Sadly, early success in the specialized field of dressage meant that in the long run this horse was then limited to a very insular life. Once taken out

Above: Take your horse on holiday and make use of the facilities or expert tutors. Remember why you bought a horse in the first place and have some fun.

focus was drowned out by everything else and he would panic. One minute he would be superb, then a bird would flap in a bush or a car would go past and he would charge blindly along the lane, oblivious to danger. Even in the company of a calm horse, it would take an unacceptably long time to calm him down.

He had jumping ability also, but such a horse could never be trusted to compete in a one-day event or even a small cross country, however low key it was. At the age of 14 a horse that should have been a schoolmaster was 'disabled' by a lack of cross-training and adequate familiarization with the world around him. It is a story from which we can all learn a lesson.

Below: Life is wonderful! Time in the company of a well-behaved horse and the bond that that experience brings is something that makes all the sweat, tears and hours of hard work suddenly seem worthwhile.

Above: Try a different style of riding and find interests that inspire you to work with the strengths of your horse, rather than always expecting him to fit in with your aspirations and dreams!

of the security of an arena or familiar surroundings, panic would set in meaning the horse became too unpredictable to ride out.

He used to be very keen to hack out but it seemed that his brain could not cope with the barrage of new sights and sounds. It almost seemed as if the volume of external messages that he was trying to assimilate all at once was just too much for him. His normal routine meant that he was principally used to focusing on the instructions from his rider; out on a hack this

■ INDEX

Note: Page numbers set in *italic* type refer to picture captions.

■ INDEX

■ PICTURE CREDITS

PICTURE CREDITS

PICTURE CREDITS

Unless otherwise credited below, all the photographs that appear in this book were taken by **Neil Sutherland** especially for Interpet Publishing.

Bitless Bridle UK: 102 bottom right.
Philip de Ste Croix: 17 bottom.
Owen Griffiths: 41 top right.
iStockphoto.com:
  AtWaG: 60 bottom right.
  Jean-Yves Benedeyt: 60 left, 116 bottom left.
  Robert Bogue: 121 bottom right.
  Roger Branch: 75.
  Dan Brandenburg: 115 bottom.
  Natasha Brownfield: 74.
  Phil Cardamone: 9 top centre.
  Ben Carlson: 110 top left.
  Cathleen Clapper: 11 bottom right, 40 left, 54 bottom left.
  Jeff Clow: 33 top left, 124 bottom right.
  Jacques Croizer: 66-7.
  Barry Crossley: 109 top right, 121 bottom centre, 141 centre right, 145 top.
  Jeff Dalton: 1.
  Derek Dammann: 59 top right.
  Dainis Derics: 20 centre right, 55.
  Andy Didyk: 93 bottom right.
  Will Evans: 113 bottom.
  Rob Friedmann: 97 left.
  Karen Givens: 112 top, 133 bottom right.
  Dieter Hawlan: 19 bottom.

Terry Healy: 144.
Johann Helgason: 14 top left, 52 bottom right, 70.
Angela Hill: 36 top right.
Andrew Howe: 90 bottom right.
Ralph Huyskens: 18 top right.
Stuart Ingram: 130 bottom.
Jamsi: 96 left.
Slawomir Jastrzebski: 114.
Don Joski: 72 bottom right.
Stefan Junger: 20 bottom right.
Linda King: 9 top right.
Anne Kitzman: 25 top.
Mikhail Kondrashov: 58 top right, 125, 149 bottom right.
Anna Kowalczyk: 2-3.
Simon Krzic: 155 bottom right.
Holly Kuchera: 4-5, 14 centre right, 103 right.
Life Journeys: 9 bottom.
Nora Litzelman: 104 left.
Michelle Malven: 92 bottom right.
Markanja: 48 left, 51 lower left, 140 both, 141 top left, 141 bottom right.
Patricia Marroquin: 49 top left.
Randy Mayes: 90 top right.
Denise McQuillen: 28 right.
Martha Moody: 33 bottom.
Sharon Morris: 51 top right.
Michel Mory: 146 top right.
Fabio Omes: 73 top left.
Alon Othnay: 142 bottom.
Marcel Pelletier: 44 bottom right.

Alexander Briel Perez: 137 bottom.
Phosan Photography: 94.
Photo Inc: 54-5.
Heiko Potthoff: 126 top right.
Piet Rijkhoff: 98-9.
Kevin Russ: 76 bottom right.
Isaac Santillan: 122 top right.
Dirk Schaefer: 6 bottom, 99 inset, 146 centre left.
Oscar Schnell: 129 top centre.
Jasson Schrock: 77 right.
Jean Schweitzer: 154 top.
Sky Creative: 110 bottom.
Lori Sparkia: 17 top.
Eline Spek: 116 centre right, 154 bottom.
Arkadiusz Stachowski: 103 left.
Michaela Steininger: 127 bottom.
Nick Stubbs: 14 centre left.
Yuriy Sukhovenko: 112 bottom.
Rob Sylvan: 33 top right.
Annamaria Szilagyi: 14 bottom.
Mikhail Tolstoy: 137 top right.
Baldur Tryggvason: 12-13.
Emrah Turudu: 72 bottom left.
Marlon van Sas: 16 left.
Don Wilkie: 14 top right.
Dawn Young: 15.
Zastavkin: 155 top left.
Jeanette Zehentmayer: 128 top.
Bob Langrish Equestrian Photographs: 6 top, 13 inset, 18 left, 76 bottom left, 78 bottom left, 150 centre right.